5G GREEN COMMUNICATION NETWORKS FOR SMART CITIES

5G GREEN COMMUNICATION NETWORKS FOR SMART CITIES

Edited by
Devasis Pradhan
Rajeswari
Hla Myo Tun
Naw Khu Say Wah
Thandar Oo

APPLE ACADEMIC PRESS

First edition published 2025

Apple Academic Press Inc.
1265 Goldenrod Circle, NE,
Palm Bay, FL 32905 USA

760 Laurentian Drive, Unit 19,
Burlington, ON L7N 0A4, CANADA

CRC Press
2385 NW Executive Center Drive,
Suite 320, Boca Raton FL 33431

4 Park Square, Milton Park,
Abingdon, Oxon, OX14 4RN UK

Library and Archives Canada Cataloguing in Publication

Title: 5G green communication networks for smart cities / edited by Devasis Pradhan, Rajeswari, Hla Myo Tun, Naw Khu Say Wah, Thandar Oo.
Names: Pradhan, Devasis, editor.
Description: First edition. | Includes bibliographical references and index.
Identifiers: Canadiana (print) 20240465407 | Canadiana (ebook) 20240465458 | ISBN 9781774917961 (hardcover) | ISBN 9781774917978 (softcover) | ISBN 9781003560357 (ebook)
Subjects: LCSH: 5G mobile communication systems—Environmental aspects. | LCSH: Smart cities.
Classification: LCC TK5103.25 .A14 2025 | DDC 004.6—dc23

Library of Congress Cataloging-in-Publication Data

CIP data on file with US Library of Congress

ISBN: 978-1-77491-796-1 (hbk)
ISBN: 978-1-77491-797-8 (pbk)
ISBN: 978-1-00356-035-7 (ebk)

About the Editors

Devasis Pradhan
Assistant Professor Grade 1 and Associate Dean of Research and Development, Acharya Institute of Technology, Bengaluru, Karnataka, India

Devasis Pradhan has been working as an Assistant Professor Grade 1 and Associate Dean of Research and Development at the Acharya Institute of Technology, Bengaluru, Karnataka, India, since 2017. His current research includes the effectiveness of 5G green communications, mmWave antenna design, UWB antennas, and their implementation. With an experience of over 16 years in academics and research, he has published more than 76 research papers and submitted eight papers *to IEEE Access*. He has authored four books and edited three books (CRC, ISTE, Bentham Science) with a reputed publishing house. He is a coeditor, editorial board member, and peer reviewer for eight international journals and is a committee member in reputed organizations. He has authored and coauthored 15 book chapters. He has been a technical committee member and reviewer for reputed international conferences, such as IEEE, Grenz Society, IFERP, etc. He has received seven national awards in the field of academic and research work from various governing bodies associated with the government of India. He is also an active member of ISTE, IEEE, ATMS, and other professional associations working toward professional growth. He has received a consultancy for technical support and testing of an RF absorber from NEOSTICK PVT Ltd.

Rajeswari, PhD
Professor and Head, Department, Electronics and Communication Engineering, Dean Academics, Acharya Institute of Technology, Bangalore, Karnataka, India

Rajeswari, PhD, is presently working as a Professor in the Department of Electronics and Communication

Engineering at Acharya Institute of Technology, Bangalore, India. She completed her PhD in the field of Speech Processing. Her areas of interest include speech processing, AI, computer vision, and applications in the fields of healthcare and agritech.

Born in a small town, Dr. Rajeswari's curiosity for the natural world was evident from an early age. She pursued her passion for science and technology and earned her bachelor's degree in Electronics and Communication from Bangalore University, MS Ramaiah Institute of Technology, Bangalore, India. She completed her ME in Electronics and Communication, Bangalore University, UVCE, Bangalore, followed by a PhD in Electrical and Electronics Engineering Sciences (area research: Speech Processing) at the Aeronautical Development Agency (ADA, Ministry of Defense, Govt. of India) at Visvesvaraya Technological University (VTU), Belgaum, India.

She formerly worked at the Acharya Institute of Technology for two decades and set an example in the field of academics and research work through imparting knowledge to countless aspiring students and researchers. Her dynamic teaching style and ability to simplify complex concepts made her a beloved figure among students. Today, Dr. Rajeswari stands as a beacon of inspiration in the Acharya Group of Institutions. Her dedication to advancing technology, mentoring future engineers, and promoting diversity has solidified her legacy as a true pioneer in the field.

Hla Myo Tun, PhD

Pro-Rector (Research), Yangon Technological University, Gyogone, Yangon, Republic of the Union of Myanmar

Hla Myo Tun, PhD, is a leading Pro-Rector for Research and Engineering, Higher Education at Yangon Technological University (YTU), Gyogone, Yangon, Republic of the Union of Myanmar. He specializes in professional training for engineering higher education leaders, heads of departments, and faculty members. He also directs research and development programs and workshops and has worked as a certified Quality Assurance Evaluator of the Myanmar Engineering Council (MEngC) since 2019.

Since 2017, he has been a project coordinator and trainer of International Postgraduate Diploma Programs in Telecommunication Engineering, which was a collaboration between the University of Oulu in Finland and

YTU. Since 2012, he has been heading the department development plans, especially in the Semiconductor Electronics Technology Research Group Establishment at Mandalay Technological University (MTU) and YTU, under the scheme of JICA Project for Enhancement of Engineering Higher Education (EEHE) in Myanmar.

Before being promoted as the Pro-Rector (Research) of YTU, Dr. Tun worked as the head of the Department of Electronic Engineering of MTU and YTU in Myanmar. He was managing two research groups on Antenna Engineering and Semiconductor Electronics Engineering Technology.

After earning his doctorate in Electronic Engineering in 2008, he headed a Control System Design Research Lab in MTU. He did his research work on Time Reversal Focusing on Underwater Communication under the Signal Processing Research Group of IIT Delhi in India in 2013. He received an award for outstanding research report from IIT Delhi. He has published over 300 peer-reviewed research journals and has attended international conferences in electronics related areas. He received a best paper award at the International Conference on Science and Engineering 2017. Dr. studied Semiconductor Device Fabrication and Properties Measurement (completing country focus training in Japanese Universities) from 2014 to 2019. Dr. Tun is an International Dean who is recognized by DAAD in Germany after completing the International Deans' Courses Southeast Asia 2022–2023. He received his Bachelor of Engineering degree in Electronics from Mandalay Technological University in 2002. His Master of Engineering Degree in Electronics is from Yangon Technological University in 2003. In 2008, he received his Doctor of Philosophy in Electronic Engineering from Mandalay Technological University.

Naw Khu Say Wah
Lecturer, Yangon Technological University (YTU), Myanmar

Naw Khu Say Wah, PhD, is currently employed as a Lecturer at Yangon Technological University, Myanmar. Prior to this, she has worked as an Assistant Lecturer at Loikaw Technological University and as a Demonstrator at Hmawbi Technological University, both in Myanmar. She has a number of conference papers to her credit as well as several journal publications in reputed national and international publications. She was awarded her PhD from Yangon Technological University, Myanmar, in Electronic Engineering while her bachelor's and master's degrees in Electronics and Communication are both from Hmawbi Technological University, Myanmar.

Thandar Oo
Tutor, Department of Electronics Engineering, Faculty of Electrical and Computer Engineering, Yangon Technological University (YTU), Myanmar

Thandar Oo, PhD, is currently working as a tutor in the Department of Electronics Engineering, Faculty of Electrical and Computer Engineering, Yangon Technological University (YTU), Myanmar. She has published several papers in journals and conferences. Dr. Oo is a dynamic tutor and researcher who is constantly working with computers to broaden her horizons. She earned her PhD in Biomedical Engineering from Prince of Songkla University, Haytai Campus, Thailand. She acquired her Master's in Electronic Engineering from Mandalay Technological University, Myanmar, and Bachelor's in Electronic Engineering from Technological University Myitkyina, Myanmar.

Contents

Contributors

Arun Agarwal
Department of ECE, ITER, Siksha 'O' Anusandhan Deemed to be University, Bhubaneswar, Odisha, India

Tarique Akhtar
Senior Data Scientist, Agility

Sunil Kumar Behera
I & C Engineering, Reliance Jio, Odisha, India

Alessandro Bruno
IULM Libera, Università di Lingue e Comunicazione di Milano, Milan, Italy
Research Department, Yangon Technological University, Republic of the Union of Myanmar, Yangon, Myanmar

Debasis Das
Department of MIS, Webster University, Tashkent, Uzbekistan

Nitin S. Goje
Department of MIS, Webster University, Tashkent, Uzbekistan

Priyadarshi Kanungo
ECE Department, C. V. Raman Global University, Bhubaneswar, Odisha, India

Dara Jogesh Kumar
HTC Global Services, Begumpet, Hyderabad, Telengana, India

N. Prasanthi Kumari
Electrical Department, SoE, UPES, Dehradun, Uttarakhand, India

Tirtha Majumder
Department of Electrical and Electronics Engineering, School of Engineering and Technology, Adamas University, Kolkata, West Bengal, India

C. S. Mala
Department of ECE, BMS Institute of Technology & Management, Bengaluru, India

Saurabh Narendra Mehta
Department of ECE, Vidyalankar Institute of Technology, Mumbai, Maharashtra, India

Gourav Misra
School of Electronic Engineering, Dublin City University, Glasnevin, Dublin, Ireland

Sasmita Mohapatra
Department of Electronics and Communication Engineering, Sir M Visvesvaraya Institute of Technology, Bengaluru, Karnataka, India

Youssef Moukabaa
ECE Department, C. V. Raman Global University, Bhubaneswar, Odisha, India

Zaw Min Naing
Department of Research and Innovation, Ministry of Science and Technology, Yangon, Myanmar

I. G. Naveen
Department of ECE, Nitte Meenakshi Institute of Technology, Bengaluru, Karnataka, India

T. Mary Neebha
ECE Department, Karunya Institute of Technology and Science, Coimbatore, Tamil Nadu, India

Thandar Oo
Department of Electronic Engineering, Yangon Technological University, Yangon, Myanmar

Smita Parija
ECE Department, C. V. Raman Global University, Bhubaneswar, Odisha, India

Devasis Pradhan
Department of Electronics and Communication Engineering, Acharya Institute of Technology, Bengaluru, Karnataka, India

Rajeswari
Department of Electronics and Communication Engineering, Acharya Institute of Technology, Bengaluru, Karnataka, India

N. R. Asha Rani
Department of Civil Engineering, Alliance College of Engineering and Design, Alliance University, Anekal, Karnataka, India

Prasanna Kumar Sahu
Department of Electrical Engineering, National Institute of Technology, Rourkela, Odisha, India

I. G. Saritha
Department of ETE, BMS Institute of Technology and Management, Bengaluru, Karnataka, India

Varun Sharma
School of Electronic Engineering, Dublin City University, Glasnevin, Dublin, Ireland

Vipul Sharma
ECE Department, FET, Gurukula Kangri (Deemed to be University), Haridwar, Uttarakhand, India

M. S. Sowmyashree
Department of ETE, BMS Institute of Technology and Management, Bengaluru, Karnataka, India

Hla Myo Tun
Research Department, Yangon Technological University, Republic of the Union of Myanmar Yangon, Myanmar

Atul Varshney
ECE Department, FET, Gurukula Kangri (Deemed to be University), Haridwar, Uttarakhand, India

Subhash Chander Verma
Data Scientist, Wipro Limited, Bengaluru, Karnataka, India

Naw Khu Say Wah
Department of Electronic Engineering, Yangon Technological University, Yangon, Myanmar

Preface

In an age where technological innovation is reshaping our world at an unprecedented pace, the convergence of 5G technology and sustainable urban development stands as a beacon of progress. Our book, *5G Green Communication Networks for Smart Cities,* serves as a testament to the transformative potential of these two intersecting realms: 5G communication networks and the pursuit of diverse, intelligent urban landscapes.

The introduction of 5G technology represents a quantum leap in the world of telecommunications. It promises lightning-fast speed, minimal latency, and a vast ecosystem of interconnected devices. However, as we ride the waves of this digital revolution, we are compelled to address the pressing challenges of environmental sustainability. Our preface illuminates the essence of this book, where we explore the symbiotic relationship between 5G technology and eco-conscious communication networks and devices with the usage of low energy to make the environment green.

The term "5G green communication network" embodies our commitment to fostering innovation that does not come at the cost of our planet. Throughout this book, we navigate the intricate strategies and solutions that allow 5G networks to operate with unprecedented efficiency, reducing their ecological footprint and embracing renewable energy sources. As our urban landscapes continue to evolve, smart cities are emerging as hubs of innovation, where cutting-edge technologies and intelligent infrastructure converge. In these cities, the quality of life for residents is paramount, and our book dives deep into the myriad ways 5G technology empowers urban development. From healthcare and education to transportation and public safety, 5G networks enable smart cities to embrace diversity in their urban planning and development.

The book chapters have been contributed by authors who are engaged in teaching, research, and extension services in their respective fields. They have incorporated their experience and knowledge along with the latest information in line with 5G communication and key enablers.

We are confident that this book will be a useful guide and resource for students, researchers, teaching faculty, and others working on 5G green communication and smart city issues.

We express our sincere gratitude to the various scientists who gave up their time and provided useful information for this book.

Introduction

In the 21st century, our world is undergoing a profound transformation driven by rapid technological advancements, environmental consciousness, and the aspiration to create more inclusive, intelligent, and sustainable urban environments. At the heart of this transformative journey lies the intersection of 5G technology, green communication networks, and the pursuit of diversity within smart cities. Our book, *5G Green Communication Networks for Smart Cities*, serves as a beacon guiding us through this convergence of technology, sustainability, and societal progress. It is a testament to the power of innovation and collaboration in shaping the future of our cities.

The advent of 5G technology has ushered in a new era of connectivity. With unprecedented speed, minimal latency, and the capacity to connect billions of devices, 5G networks promise to redefine how we communicate, work, and live. However, as we embrace the possibilities of this digital frontier, we must also confront the pressing challenge of environmental sustainability. The concept of a 5G green communication network encapsulates the idea that technological progress and ecological responsibility can coexist harmoniously. In this book, we embark on a journey to explore the intricate strategies and innovations that allow 5G networks to operate efficiently while minimizing their carbon footprint.

The book discusses various aspects of 5G components, such as 5G technologies, green communication, antenna design for low utilization of energy for communication process, and semiconductor devices used as key enablers. The book also focuses on network protocols for the health sector and monitoring the critical issue in order to maintain electronic health records with the amalgam of blockchain technologies. It also covers the industrial aspect of Industry 4.0 for supply chain management, edge computing, and communication between device-to-device or machine-to-machine. Our exploration takes us beyond the realms of connectivity to consider the profound impact that 5G green communication networks have on the development of smart cities.

This book is an invitation to envision a future where technology is the enabler, sustainability is the guiding principle, and diversity is the cornerstone of our cities. As we embark on this intellectual journey through the pages that follow, we invite you to explore the transformative potential of 5G green communication network and diversity for smart cities and join us in shaping a world where our urban landscapes are not only smart but also green, inclusive, and responsive to the diverse tapestry of human life.

Novel figures of excellent quality are labeled in the book to support the assembled chapters with proper citations by the contributed authors. In addition, appropriate and current references are provided in each chapter. Data given on various aspects of 5G key enablers to various dimensions of smart cities will be useful for academicians and analysts. In addition, the book provides sufficient information for college, university undergraduate, and graduate studies. It will also give a better dimension of exploration to people in industrial management.

Acknowledgment

We express our gratitude to all the contributors to this book, whose expertise and insights have helped to shape and enrich the discussions on 5G green communication networks and diversity for smart cities.

We also thank our families, friends, and colleagues for their support and encouragement throughout this project.

We acknowledge the support of our respective institutions and organizations, which provided us with the resources and infrastructure necessary to undertake this work.

Finally, we want to express our appreciation to the readers of this book, for their interest in and commitment toward advancing the field of 5G and sustainability. We hope that this book will serve as a valuable resource and inspiration for researchers, practitioners, and policymakers working in this area.

CHAPTER 1

Role of 5G Network in Smart Cities Toward Green Technology

YOUSSEF MOUKABAA, SMITA PARIJA, and PRIYADARSHI KANUNGO

ECE Department, C. V. Raman Global University, Bhubaneswar, India

ABSTRACT

This paper explores the pivotal role of 5G technology in smart cities, emphasizing its contributions to green technology. With ultra-low latency, massive device connectivity, and high-speed data transfer, 5G enables innovative solutions for urban challenges like traffic management, energy consumption, and environmental monitoring. Key applications include intelligent transportation systems, smart grids, and IoT-based environmental sensors, all contributing to reduced carbon footprints and improved resource efficiency. By facilitating real-time data collection and analysis, 5G supports optimized energy use and reduced emissions, while promoting renewable energy deployment and sustainable urban planning. The study highlights the necessity of strategic planning and stakeholder collaboration to address infrastructure, cybersecurity, and regulatory challenges. Ultimately, 5G technology is identified as a critical enabler for achieving sustainability goals in smart cities, driving progress toward greener, more resilient urban environments.

1.1 INTRODUCTION

The implementation of 5G networks actually began in 2019, when "63 operators in 35 countries launched one or more 3GPP-compliant 5G

5G Green Communication Networks for Smart Cities. Devasis Pradhan, Rajeswari, Hla Myo Tun, Naw Khu Say Wah, & Thandar Oo (Eds.)

services (including 55 mobile and 34 FWA services)".[1] This was driven by the increase of data demand and the need to minimize the latency to adapt to new use cases like driverless cars, augmented reality, healthcare applications, and IoT that are not satisfied by the current LTE networks. According to the latest Ericsson Mobility Report,[2] Mobile traffic is expected to grow by 27% annually between 2019 and 2025. Continuing recent trends, most of this will come from video traffic. Global total mobile data traffic is expected to grow by a factor of 4 to reach 160 exabytes per month in 2025, and 5G networks will carry around half of it.

The 5G network should address this increase in traffic demand and the growth of connected devices, and according to Ref. [2], the 5G networks should support the following:

- More than 1 Gbps cell throughput and a cell edge rate not less than 50 Mbps.
- Ultralow latency, preferably less than 10 ms, and even around 1 ms for some applications like tactile Internet and some industrial applications.
- Ultrahigh reliability and availability and user consistency.
- Supporting machine-type communication (MTC) devices, which consume very little power and are very energy efficient.

In order to satisfy all these requirements, many challenges have to be addressed, and some of them, like low latency, may require redesigning the core network and an upgradation of the hardware and the electronics used.

Another main concern of 5G development was reducing power consumption. However, these requirements created a conflict. Higher throughput would mean moving toward mmWave and massive MIMO. Nevertheless, the usage of mmWave means a smaller coverage area for each Base Station (BS). This means that to provide full coverage, the number of BSs should increase, creating a higher energy-consuming and polluting network. Therefore, researchers focused on mitigating the energy consumption while ensuring that the main Key Performance Indicators (KPIs) are still above the 5G threshold. In this work, we discuss several suggested methods to enhance power saving and compare them. Moreover, we performed several simulations to study and compare the performance of both 4G and 5G networks regarding latency, throughput, and energy consumption under different scenarios and illustrated how 5G could be the key to green technology. The results demonstrate that 5G can provide enormous enhancements in all these fields.

1.2 ENERGY ISSUES IN 5G AND PROPOSALS

The energy issues should be considered from both the network (specifically gNBs) and the User Equipments (UEs). What we are more concerned about in our chapter is the network's point of view since it is the more power-consuming and environment-polluting part.

1.2.1 BASE STATION ENERGY CONSUMPTION

The majority of the energy consumed by mobile networks is being consumed by Base Stations (BSs). Therefore, optimizing the power consumption of 5G networks should focus on BSs. However, 75–90% of the time, these BS resources are not used, even in highly congested networks.[3] The main reason for this is that the BS components remain operational all the time, so they can transmit necessary idle messages such as reference signals, synchronization signals, and system information.

According to Huawei, a 5G Base Band Unit/Radio Receiver Unit (BBU/RRU) needs more than 11.5 kW of power, around 70% more than a base station deploying 2G, 3G, and 4G radios.

To overcome the need for extra power in 5G, a number of advancements were made. In the following subsections, we are going to talk about two main power reduction contributors: beamforming and deep sleep.

1.2.1.1 ADVANCED "DEEP SLEEP" MODE

The basic principle of Deep Sleep is to turn off one or more devices of the base station selectively when there is no traffic. This feature is limited to the 4G network due to its interface design.[5,6]

1.2.1.2 BEAMFORMING AND MASSIVE MIMO

The bandwidth that the 5G antennas manage is five times higher than in 4G. Hence, it can deliver higher throughput and serve more users using massive MIMO. This means that the required time to download or upload a file is reduced significantly, leading to less power consumption.

Using beamforming increases data rates, channel efficiency and reduces interference. Therefore, the wireless protocols can calculate the minimum energy needed to deliver the data, leading to less power consumption for both the user equipment and the base station. Therefore, 5G networks that use beamforming consume four times less energy than 4G ones.[5–7]

1.2.2 ENERGY CONSUMPTION IN MMWAVES

An important feature of 5G is that it allows deployment over a wide range of frequencies from 1 GHz to over 50 GHz. Therefore, it is divided into two ranges: below 7 GHz (FR1) and above 7 GHz (FR2).

Control channel monitoring in FR2 consumes 75% more energy than in FR1, as can be seen in Table 1.1, but since the transmission of data can be completed in less time, the required energy to transmit a data burst of a given size using FR2 is less than that in FR1.

TABLE 1.1 Device Power Consumption in Different Operations and States (TR 38.840).

Power consumption (relative units)		
Device power states and operations	**FR1 (below 7 GHz)**	**FR2 (mmWave)**
Deep sleep	1	1
Light sleep	20	20
Microsleep	45	45
PDCCH monitoring only	100	175
SSB measurements	100	175
CSI-RS measurements	100	175
PDCCH+PDSCH reception	300	350
Uplink transmission (depends on TX power level)	250–700	350

1.3 PERFORMANCE EVALUATION FOR ONE USER WITH MULTIPLE SPEEDS

As mentioned in previous sections, 5G has overcome the weaknesses and enhanced the overall performance of LTE networks, especially in terms of throughput and latency. The development of 5G is mainly driven by the need for ultralow latency and high throughput for online gaming, virtual reality, and 3D videos. However, the usage and deployment of 5G mmWave systems introduced several difficulties and issues that made moving toward relying on them solely very difficult and impossible in the next few years. Therefore, dual connectivity with LTE networks was introduced.

Our suggestion is to use the Fast Switching (FS) DC in future networks but with the difference of making the LTE network the default and preferred network for all UEs. Our suggestion is based on the following facts: (i) the LTE networks around the world are already implemented and mature, and their availability is more than 80% in most countries; (ii) the throughput and download speed provided by LTE are sufficient for many applications that are currently used and will still be used in the future (browsing, instant messaging, and image and video sharing). In the following sections, we will provide some statistics on the achievable rates of LTE and the results of a simulationwe conducted.

1.3.1 REAL LTE STATISTICS

The following statistics are taken from a report published by OpenSignal.[1] The measurements in this report were conducted between January 1, 2019 and March 31, 2019 using 43,614,234 devices, and a total of 139,960,248,468 measurements were taken from 87 countries.[8]

1.3.2 DOWNLOAD SPEED

From Figure 1.1, we can notice that the download speed for LTE varies enormously among countries, with the best value above 50 Mbps, and the average score of the 87 countries tested is 17.6 Mbps. These values are more than enough for even UHD videos, which require a download speed of around 5 Mbps.

---

[1]Opensignal is the independent global standard for analyzing consumer mobile experience.

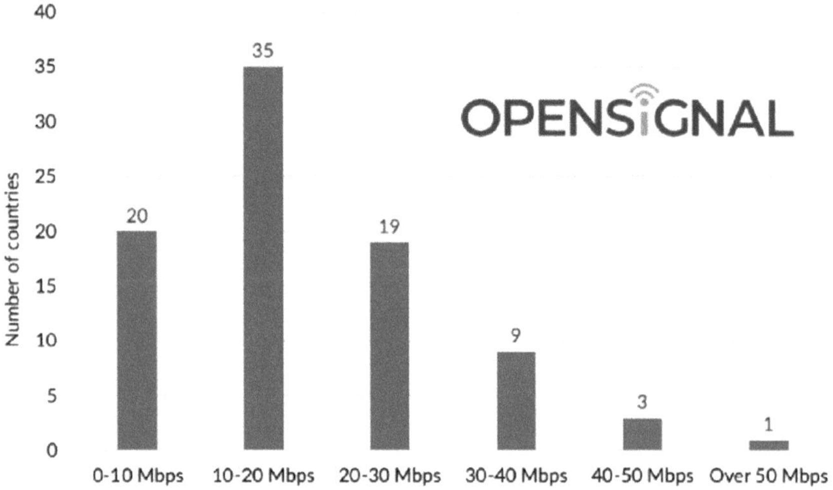

FIGURE 1.1 The download speed range distribution.

Source: Reprinted from Ref. [8]

1.3.3 *LATENCY*

In terms of latency, no country managed to achieve less than 30 ms latency, and the average was under 40 ms. This means that LTE is not sufficient for real-time and V2V applications. The results of each country are illustrated in Figure 1.2.

FIGURE 1.2 The latency in the studied countries.

Source: Reprinted from Ref. [8]

1.3.4 SIMULATION

The simulation conducted by us aims to compare the values of latency and throughput of both LTE and 5G under the same scenario (UE location, UE speed, building distribution, start and end points, etc.). For this, we based our work on the *mc-twoenbs.cc* example, which can be found in src/mmWave examples. Originally, this example instantiates an LTE and two mmWave eNBs and attaches one UE to both and starts a ow for the UE to and from a remote host.

We modified this example to simulate the performance of an LTE network only and a 5G DC network; the LTE used is the standard LTE without using the DC configuration, and we used the default parameters. The main parameters can be found in Table 1.2.

We conducted this simulation on a PC using a core i5 @2.80 GHz processor, 16GB of RAM, and operated by Ubuntu 18.04. The simulation scenario parameters are as follows:

- The UE start point (80, −5, 1.6);
- The UE end point (120, −5, 1.6);
- Buildings: 8 buildings with a random distribution.

We conducted the simulation multiple times, where we changed either the download data rate or the UE speed, to see the saturation point of each technology and the effects of the UE speed on the latency and throughput.

TABLE 1.2 Main Parameters.

Parameter name	5G simulation value	4G simulation value
Access technology	OFDMA	OFDMA
Duplex	TDD	FDD
Bandwidth	1 GHz	25 MHz
Number of UE antennas	16	1
Number of Enb antennas	64	1
Number of UEs	1	1
Number of LTE Enbs	1	1
Number of mmWave Enbs	2	None
Velocity model	Constant	Constant
Data rate	1 Gbps	1 Gbps

1.3.5 SIMULATION RESULTS

We have done 10 different simulations, changing the positions of the eNBs and the speed of the UE, to study the effect of different speeds on the throughput, and we came up with the results illustrated in Table 1.3.

The speeds ranged from 1 m/s, which simulates indoor movement or almost stability, to 30 m/s (108 km), which represents car movements on highways. From the table, we can see that the LTE network throughput is affected by the UE speed, and the relation is inversely proportional (the faster the UE, the smaller the throughput). The overall variation is around 7 Mbps (9%). This returns to the fact that the coverage area of the LTE is larger, which means fewer handovers, and to the fact that the LTE radio channel is more stable than the mmWave one. While in the 5G mmWave, the effect of the speed is bigger and the relation is not purely inversely proportional, this can be explained by the mmWave radio channel vulnerability and rapid quality variations.

TABLE 1.3 LTE and 5G Maximum Throughput Values.

UE speed (m/s)	Max 4G throughput (Mbps)	Max 5G DC throughput (Mbps)
1	74.77	381.74
2	74.33	416.55
3	73.91	407.79
5	73.16	320.18
10	71.59	397.79
15	70.27	359.71
20	69.33	375.42
25	68.48	222.85
30	67.75	212.38

The maximum throughput varied between 416.55 and 212.38 Mbps, which means the throughput dropped by around 50%.

We wanted to see the effect of both the relative distance between the UE and eNBs and different building distributions solely without the effect of movement speed.

In the first scenario, the first mmWave gNB is located at points (5,70), and its height is 3, while the second mmWave gNB is located at points

(150,70) and its height is also 3. In the second scenario, the coordinates of the first mmWave gNB are (5,70), and its height is 3, while the seconds are (170,70) and its height is also 3. In both scenarios, the LTE eNB is colocated with the first gNB. The UE movement starts at (80,70) and ends at (120,70), and its height is 1.6. When we change the locations of the eNBs, the whole building can be changed since we use random distribution. The results of this simulation are shown in Figure 1.3.

We can easily notice that the LTE values are almost the same in both scenarios, while the 5G DC values in the second scenario are much lower than the ones in the first one, noting that the UE in the second scenario is further from the base stations than the case in the first one. These results come from the fact that the coverage area of the LTE eNB is larger than that of the mmWave gNB and also from the fact that mmWave signals are more susceptible to blockage, which also shows the importance of the signal strength and quality to maintain the high performance of the mmWave network.

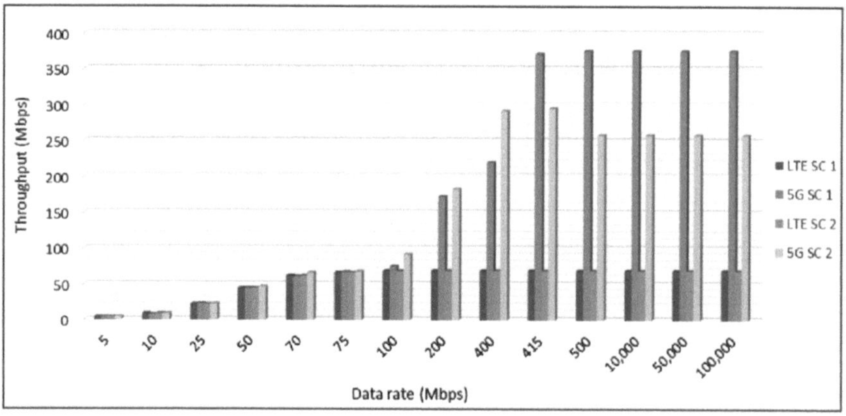

FIGURE 1.3 The throughput of both LTE and 5G DC networks in two scenarios using the same UE speeds and different BS distributions.

Overall throughput: Our main aim of the simulation is to compare the performance of the LTE and the mmWave 5G under the same circumstances. After getting the results and averaging them, we got the results illustrated in Figure 1.4 and Table 1.4.

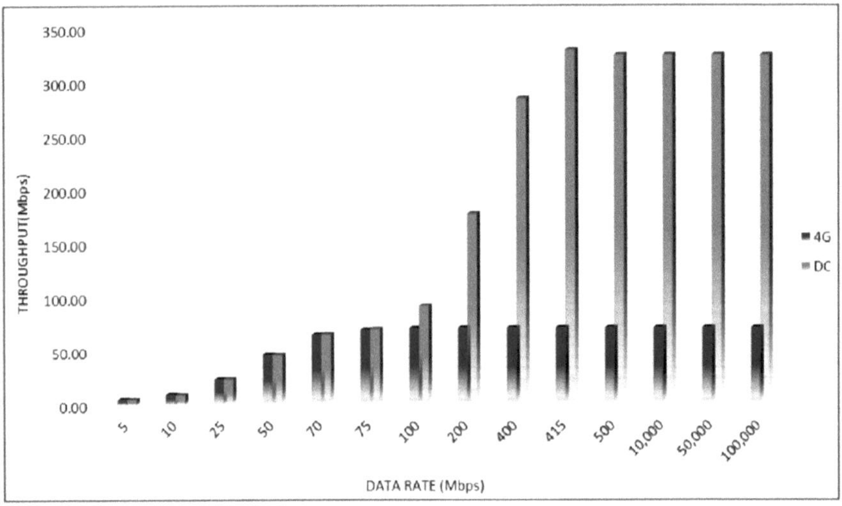

FIGURE 1.4 The average throughput of both LTE and 5G DC networks under different data rates.

TABLE 1.4 LTE and 5G Maximum Latency Values.

Mbps	4G	5G DC
5	3,499,925	4,534,812
10	3,499,884	41,310,777
25	3,500,135	25,225,352
50	3,533,036	2,798,737
70	3,536,302	3,965,737
75	3,601,796	6,348,963
100	868,389,580	52,404,041
200	1,393,512,805	191,215,889

So, in both cases (throughput and latency), the turning point is the 75 Mbps data rate since after that, the 5G network performance is much better in both cases. This comes from the fact that the maximum data rate of the LTE network with the configurations we used is 75 Mbps.

From the results, we can easily notice that the 5G network outweighs 4G by many folds sometimes.

1.4 PERFORMANCE EVALUATION FOR MULTIPLE USERS WITH MULTIPLE SPEEDS

In the previous section, we simulated the performance of both LTE and 5G mmWave DC networks using one UE. Despite the importance of this simulation to see the limits and ideal performance of the networks, it does not give us a realistic indicator of their performance.

Many factors are not taken into consideration when simulating for one UE only, like self-interference and the high rate of data and control messages, as well as the effects on transportation parts.

In this section, we will perform a simulation using different numbers of UEs to examine the impact of increasing the number of UEs on the throughput and latency.

The simulation parameters are similar to those in Chapter 5, with the main difference being in the number of UEs used.

1.4.1 SIMULATION

We simulated different numbers of users (from 1 to 50 UEs) under the same scenarios in both networks to see how each network will act and the effects on the network and on each individual UE. We selected the 50 UEs to be the maximum number of UEs because we believe that in actual networks, this number represents the maximum number of UEs that could be active simultaneously in this size of an area. Based on the results of the previous simulation, we selected the data rate to be 1 Gbps. Moreover, since we saw the effects of mobility on the performance, in this simulation we simulated the high and low mobility cases to see if similar effects would occur.

1.4.2 SIMULATION RESULTS

Here also, we will compare the throughput and latency of both networks and check the effects of the number of users and UE speed.

Low mobility: For the low mobility, the 1 m/s speed was used, and we got the results for both latency and throughput. The results are as follows:

The overall performance is similar to the case of one user in very high data rate cases (Figure 1.5), where the 5G network outperforms the LTE in both latency and throughput.

Here, it is actually more obvious since the 5G network throughput reached much higher levels (double) than in the case of one user. This is due to the fact that the network resources and capacity are divided among the number of users, so any limitation from the UE side will not affect the whole network performance. From Figure 1.6, we can see that the latency of the LTE network became much worse because the LTE network was not designed to handle this amount of data.

High mobility case: In the high mobility case, we used 25m/s. The throughput results are demonstrated in Figure 1.7, and the latency is shown in Figure 1.8.

From Figures 1.7 and 1.8, we can see that the 5G superiority above the LTE network is also obvious in terms of latency and throughput.

In cases of a high number of users (5 and more), we can see from the four figures that there is no direct relation between the number of users and the overall latency and throughput of the two networks. In the LTE network, the overall performance is stable in all cases, with almost an overall difference of about 0.5 Mbps between the best and worst cases in throughput and around 500 ms in latency, while in 5G, the performance fluctuates randomly in both the latency and throughput, which can be explained by the radio channel vulnerability for any obstacle.

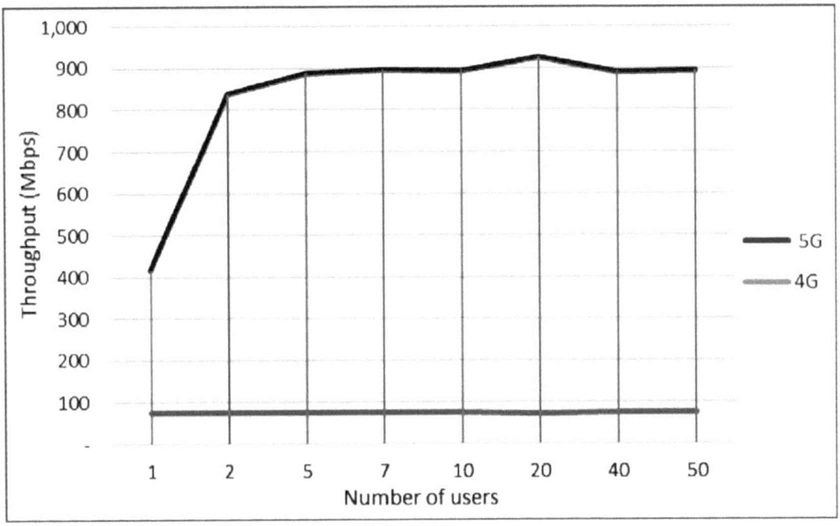

FIGURE 1.5 The throughput of both LTE and 5G DC networks for different users (speed 1m/s).

In conclusion, the superiority of the 5G network is obvious in both terms (throughput and latency), and this is not surprising since a lot of factors and enablers were used to make the 5G network much better in all terms.

1.5 CONCLUSIONS AND FUTURE WORK

The 5G network proved its superiority above the LTE network and outperformed it in terms of throughput and latency, especially in the case of multiple users. But, in some cases, the 5G network consumed more energy than the 4G network since the higher frequency would require more energy. However, when comparing the overall delivered data with the energy needed to deliver it, the 5G network needed much less wattage per bit. Hence, the energy consumption is relatively lower.

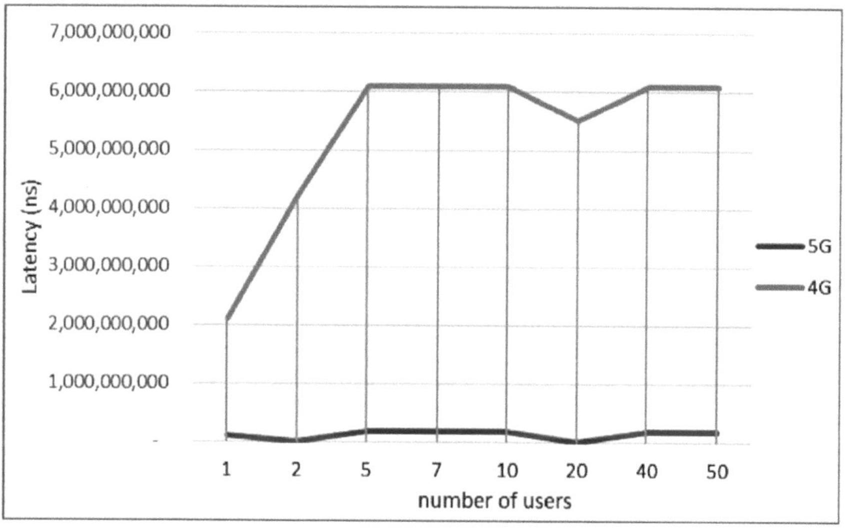

FIGURE 1.6 The delay of both LTE and 5G DC networks for different users (speed 1m/s).

In the future, we plan to perform a deeper comparison by comparing the 5G network in different modes (IoT, eMBB, and others) with each other and with 4G. Moreover, after the implementation of 5G in the real world, we would be able to collect real data under real transmission circumstances.

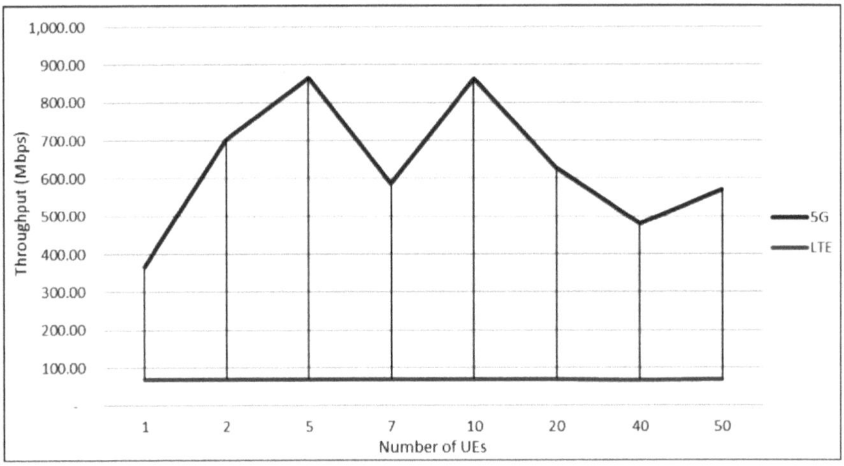

FIGURE 1.7 The throughput of both LTE and 5G DC networks for different users.

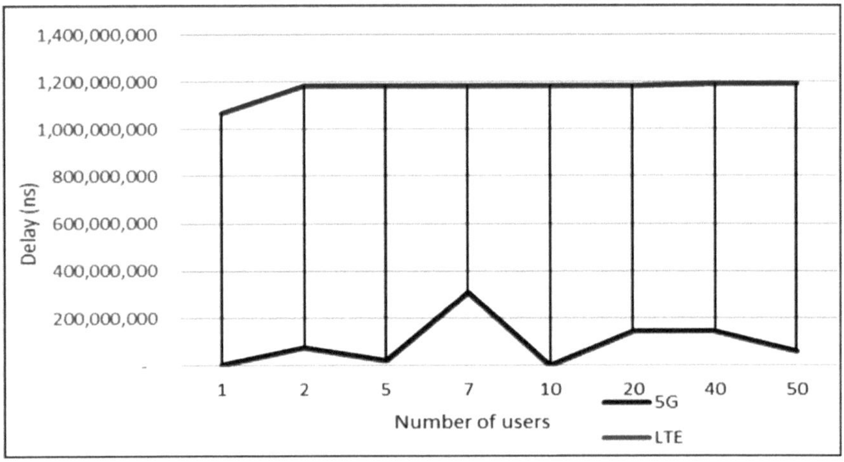

FIGURE 1.8 The delay of both LTE and 5G DC networks for different users.

KEYWORDS

- **5G technology**
- **smart cities**
- **green technology**
- **sustainability**
- **IoT (Internet of Things)**
- **energy efficiency**
- **urban planning**

REFERENCES

1. Lte and 5g Market Statistics: Global Snapshot. https://gsacom.com/technology/5g/ (accessed on Mar 2020).
2. Ericsson Mobility Report. https://ericsson.com/mobility-report (accessed on Nov 2019).
3. Shurdi, O. et al. 5G Energy Efficiency Overview. *Eur. Sci. J.* **2021,** *17*(3), 315–327.
4. Dongxu, C. 5G Power: Creating a Green Grid That Slashes Costs, Emissions & Energy Use. https://www.huawei.com/us/technology-insights/publications/huawei-tech/89/ 5g-power-green-grid-slashes-costs-emissions-energy-use
5. https://www.ericsson.com/en/blog/2020/2/mobile-devices-and-energy-efficiency
6. https://www.ericsson.com/en/blog/2019/9/energy-consumption-5g-nr
7. Nokia 5G Energy Efficiency White Paper. https://gsacom.com/paper/5g-network-energy-efficiency-nokia-white-paper/
8. Boyland, P. The State of Mobile Network Experience May 2019 Benchmarking Mobile on the Eve of the 5g Revolution. https://www.opensignal.com/reports/2019/05/global-state-of-the-mobile-network (accessed on May 2019).

CHAPTER 2

Green Communication in Next-Generation 5G Network: Techniques and Challenges.

TIRTHA MAJUMDER[1] and PRASANNA KUMAR SAHU[2]

[1]*Department of Electrical and Electronics Engineering, School of Engineering and Technology, Adamas University, Kolkata, India*

[2]*Department of Elctrical Engineering, National Institute of Technology, Rourkela, Odisha, India*

ABSTRACT

As technology advances, industry and researchers are concentrating on making communication as environmental friendly as feasible. With the advancement of the modern world, recent mobile technology concepts include an increase in the number of devices per year. The advancement of 5G in cellular technology is the next stage in meeting customer needs. 5G is going to be a dominant paradigm of communication for various emerging smart services in almost every field, such as smart cities, consumer goods, smart home, health care, manufacturing, and transportation. This is accompanied by an increase in cellular network energy usage and necessitates the need to innovate in the field of energy- efficient communications. Higher energy consumption results in increased carbon dioxide emissions and harmful radiation into the atmosphere. GREEN communication is a key technique for mitigating the ecological and health risks linked with rising carbon dioxide levels. The goal of GREEN communication is to communicate in a way that is environmental-friendly by limiting carbon

5G Green Communication Networks for Smart Cities. Devasis Pradhan, Rajeswari, Hla Myo Tun, Naw Khu Say Wah, & Thandar Oo (Eds.)

dioxide emissions. The network scenarios where energy-efficient techniques of GREEN communication can be used for better performance are device-to-device communication (D2D), massive multiple-input multiple-output (MIMO) systems, heterogeneous networks (HetNets), and the GREEN Internet of Things. This chapter aims to highlight the most recent research techniques in GREEN communications and networking for next-generation networks, as well as the challenges that come with them.

2.1 INTRODUCTION

The improvements in 4th Generation (4G) networking have motivated experts to focus on the 5th Generation (5G) network, which will enable global wireless network connectivity. Therefore, future days will be characterized by huge networks, enormous bandwidths, and unfathomably high carrier frequencies. The 5G networking will be able to support a phenomenal increase in device capacity.[1–6] Table 2.1, depicts the required specification for 5G.

TABLE 2.1 Range of Minimum Expected Requirements in 5G.

Specifications	Expectation
End-to-end latency	5 ms
Reliability	99.999
Number of devices	1M km^2
Mobility	500 km/h
Peak data rates	10 Gb/s
Mobile data volume	10 Tb/s/Km^2
Service time deployment	90 minutes

The enormous creative progress, however, is accompanied by an increase in greenhouse gases and environmental deterioration. The expansion of the 5G network is crucial, but it must be done with the least amount of energy possible. Figure 2.1, shows the overall carbon footprint in a cellular environment. The rapid increase in the number of users, devices, and applications during the past 10 years has led to an exponential growth in energy consumption levels. Increased energy consumption leads to more carbon dioxide (CO_2) being released into the atmosphere, thereby more people being exposed to harmful radiation. The telecom department emits 3% of all CO_2 emissions, which may seem like a modest amount but might be quite dangerous. With the progress of mobile technology evolution, the amount of CO_2 being emitted has increased, which has a significant impact on the global climate[7] and thereby increased global temperatures. GREEN communication is a crucial tool for reducing the threats to the environment and human health brought on by rising CO_2 levels. By reducing CO_2 emissions, GREEN communication aims to communicate in a way that is environmentally sustainable. Although the telecommunications sector is expanding swiftly, battery innovation is moving along at a comparatively slower rate. Advanced cell batteries can get through the main impediment to communications that are more energy-efficient and low power.

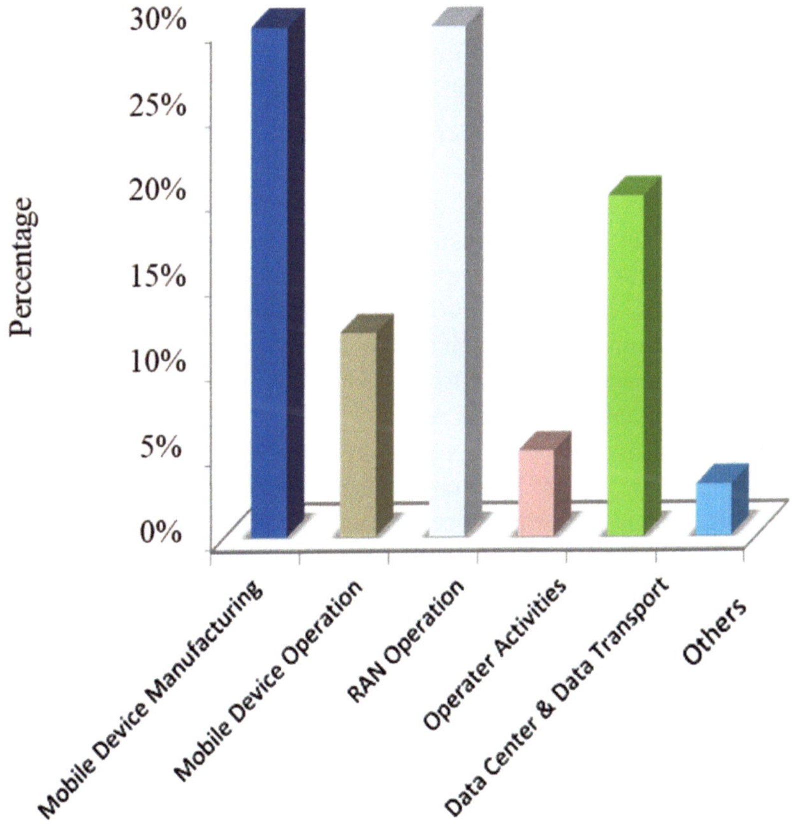

FIGURE 2.1 Carbon footprint in mobile communication.

In order to maintain control over the use of power, the following inspection areas need more focus:

- Resource Allotment: The most effective resource allocation is a significant factor influencing power utilization. The efficient use of resources in terms of time, frequency, and space has been examined. In the literature,[8–10] a variety of resource allocation techniques that can maximize energy efficiency (EE) have been studied.
- Optimal network planning: Techniques to arrange the network in energy productive way comprehends minimization in Base Stations (BS)[11] and involving sleep modes for BSs. This strategies are essential to aid less CO_2 emission.
- Renewable Energy: As BS are powered by fuel and therefore contributing to increase of CO_2 level. The use of renewable sources will be an intelligent choice to power the BSs.[12] A mobile network that is enabled by GREEN energy is an exigency. The optimization and design strategy of such a network can be seen in Ref. [13].

Energy harvesting (EH) has made wireless device powering possible.[13] It needs the accessibility of devoted sources of energy, consequently supporting extra power for utilization.[14,15] EH network deployment requires careful hardware inspection.[14] Around the world, efforts for Energy-Efficient Next-Generation Cellular Networks have been initiated. Research on 5G innovations has accelerated as a result of the evolving 4G regulations. Numerous research and study groups are conducting research into EE in next-generation networks (NGNs) on a global scale. The International Telecommunication Union dedication to GREEN Communication is impressive. Go GREEN remains its main objective, claims Study Group 5 (SG5).[15] In 3rd Generation (3G) Partnership Project Release 13, many 5G technologies have been covered. The potential energy savings of 5G have been examined in Releases 14 and 15.[16–18] The European Telecommunications Standards Institute also focuses on cellular networks with low energy consumption.[19] The 5GPP has started a variety of projects to develop 5G networks.[20]

2.1.1 CONTRIBUTION

Numerous studies have been done to improve energy efficiency (EE) and move ICT toward GREEN Communication.[16–19] A thorough analysis of 5G networks has been provided in Ref. [20]. Instead of increasing transmission powers, new approaches to network design and operation can be used to avoid the energy constraints in the networks. The network EE is intended to be increased 1000 times with such strategies. The state-of-the-art has been presented in Refs. [19, 21] for different similar techniques. In Ref. [17], a variety of techniques for power efficiency enhancement in wireless networks mainly concentrating on small cells and relays to achieve the goal of GREEN 5G Networks. The authors in Ref. [22] make a substantial addition to the GREEN Networking by thoroughly examining many projects, ongoing activities, and technology for EE improvement. The authors of Ref. [22] contribution to the GREEN Networks, was an exhaustive study on the technologies that were going on along with advances for the improved proficiency of energy. Despite the state-of-the-art, the current study outcomes are still in the early stages and face many obstacles. In addition to network EE, prolong battery life is a crucial issue that needs to be addressed.

In this paper, a thorough analysis of cellular networks those are energy-efficient in the name of environmentally friendly communication have been taken care. The rising customers demand for services has environmental impact. Therefore, power optimization at the BS is studied in detail. Numerous EE metrics have been assessed in order to analyze the network. Furthermore, various 5G technologies that might improve the EE of future networks have been researched, including massive MIMO, spectrum sharing, UDNs, IoT, and device-to-device (D2D) communication. All of these technologies have been the subject of thorough literature reviews. In order to handle the spectrum-sharing issue and achieve a prolonged battery life, critical thought must be given to this aspect. Significant research obstacles for a GREEN network are also listed.

2.1.2 ORGANIZATION

The rest of the information is coordinated accordingly: In this work Section 2.2 discusses on how raising mobile user numbers and their high data rate demand contributes to increasing CO_2 emission. Section 2.3 is highlighted with EE at the BS. Section 2.4 reviews the various energy efficiency indicators and briefly reviews power allocation approaches. Following that, various technologies that can assist GREEN networking in 5G networks are examined in Section 2.5. In Section 2.6, several related research obstacles were covered. In Section 2.7, the chapter eventually comes to an end with the conclusion.

2.2 GREEN COMMUNICATION: THE EVOLUTION

In today's world, every person is a mobile phone user. The wireless communication network is developing quickly in order to provide support to each individual client. Due to the increased traffic, the production of mobile devices and the running of the radio access network (RAN) both significantly contribute to the carbon footprint. Concern over EE has increased as a result of the escalating data rates. An economic motivation for reducing the carbon footprint internationally is to reduce the energy usage. The graph in Figure 2.2 shows how the relationship of CO_2 levels increases over time as wireless communication traffic increased.

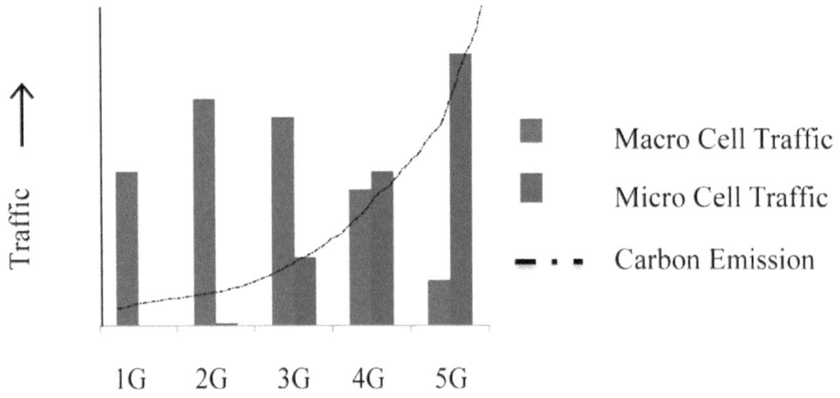

FIGURE 2.2 Traffic trends and carbon emission.

After the 3G, an effort to reduce cell size was encouraged, leading to a decrease in macro cell traffic and an increase in small cell traffic. Micro cell deployment was then started. Small cells are more practical for managing traffic indoors. The energy consumption in RAN increases as a result of providing service to the increased rate of customers. For greener next-generation networks (NGN), it is obvious that the EE of the networks needs to be improved. Table 2.2 examines the different features of 4G and 5G.

To make a clear comparison between the various wireless communication generations, the Table 2.2 takes into account a variety of elements. It is anticipated that the next-generation (5G) of networks will significantly increase CO_2 emissions. The alarming rate increase of carbon footprint is directly related to the enormous subscription needs. This focuses on the need for GREEN Communication.[23] There must be a reduction in carbon emissions. The BS's power optimization is covered first.

TABLE 2.2 Different Generation in Cellular Environment.

Feature	4G	5G
Base station density	8–10 BS/km^2	40–50 BS/km^2
Antennas in base station	8	Upto 100
Antenna type	Patch and slot	Phased array
Average power consumption/site	1.3 KW	1.1 KW
Carbon footprint per mobile subscription	23 kg	31 kg

2.3 BASE STATION POWER OPTIMIZING

In order to guarantee ubiquitous coverage, the subscription provider aim to reach out to extremely remote sites. More deployment will occur in rural areas, leading to a large rise in the number of BSs. As cellular networks draw a significant amount of their energy from BSs therefore the BSs EE enhancement is a dire need. Figure 2.3 portrays the power consumed by each parts in the BS. The energy used by the BS in the network accounts for around 80% of the total energy consumption.[23] Power amplifiers and air conditioners in the BS constantly run, absorbing up to 70% of its energy. Power amplifiers that use less energy have also been studied in Ref. [24]. Whenever a BS is active, its utilization of power is dependent on many parameters and thereby contributes significant rise in power levels. This adds the increase of CO_2 concentration in the atmosphere.

Effective power management at the BS is listed by the researchers in Ref. [25]. Power savings are further boosted by deactivating components while not in use.[26] The authors of Ref. [27] introduced TANGO, which stands for traffic-aware network planning and green operation. It is a useful framework for increasing the EE of cellular networks while maintaining network QoS. In the GREEN Communication Network, the installed new

BS equipment consumes up to 50% lesser energy. Additionally, the BSs that were diesel-powered are now using solar batteries and other forms of alternative energy. This helps in reducing the Fuel transportation expenses. With the ability to sustain three times much traffic, GREEN Communication technologies can help keep the ecological and financial systems in balance. The effective BS design is a crucial factor to understand the environmental impact of advance cellular network. Multiple GREEN BS sites include Flexi BS, Software define radio (SDR) BS, and others are listed in Table 2.3.

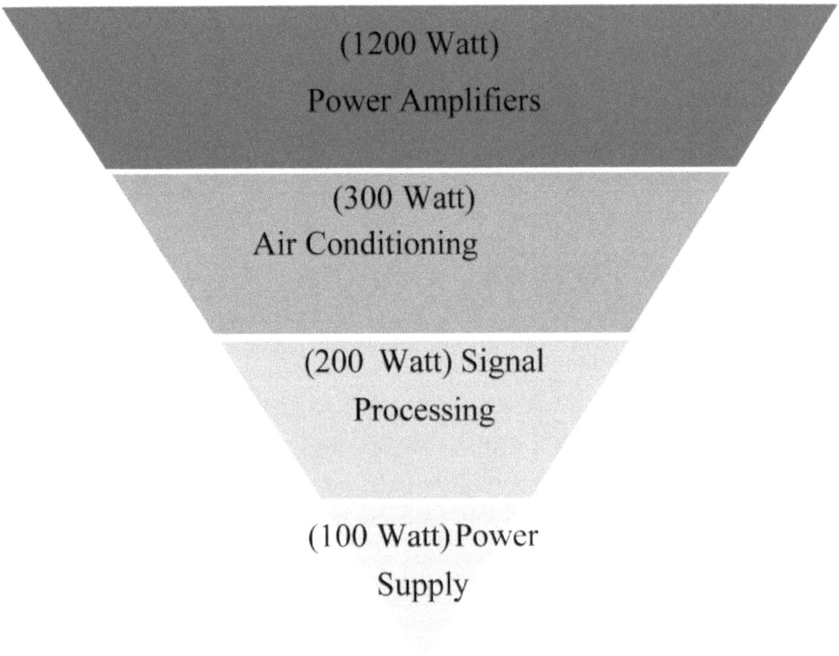

FIGURE 2.3 Power use of BS components.

Deployment of such GREEN BS sites can be beneficial to the operators for cost and maintenance charges reduction by 50%. GREEN technologies are introduced and globally accepted by Nokia Siemens Networks, Ericsson, Nujira, and other companies. Additionally, to optimize use of BS power, traffic patterns are also monitored. The various traffic patterns are determined by user activity.

TABLE 2.3 Green Communication Base Station Sites.

BS of Green Communication	Significant features
Flexi BaseStation	• Being remotely controllable.
	• Encourage the use of renewableenergy.
	• Upgradeable support for newwireless technologies.
SDR Soft Basestation	• Reduction in power consumption
	• Reduction in elements
	• Programmable
Capsule Site	• Highly suitable for installation in urban areas
Air Scale BaseStation	• Energy efficiency enhancement
	• up to 60%
	• Low cost
5G New Radio	• Ultra-low power
	• Low latency
	• high transmission rate

Traffic at daytime is more significant than night time traffic. On weekends and during holidays, less traffic persist. The peak hour, when traffic is heavy, only lasts for a brief time each day. This has been shown in Figure 2.4 for clear understanding. The coverage areas, user mobility, and kind of applications are other factors that affect traffic patterns. As the majority of data, audio, and video traffic is sporadic in nature, and its dynamic nature uses more energy,[28] it is anticipated that this pattern will persist in NGNs as well. The survey done in Ref. [29], shows the change in traffic patterns in the upcoming years. Traffic during peak hours will shortly surpass 702 Mbps.[30] It is evident that there has been a tremendous rise in traffic. In Ref. [31], when the traffic load falls below a threshold and lasts for a specific amount of time, the BSs have a tendency to enable sleep mode or turn off. During the sleep mode, the BS use a constant power, shown by the symbol P_{sleep}. BS has a different power usage[32] during the light sleep mode or in deep sleep mode. Since BSs are spread throughout a unit area, the total power consumption is given as:

$$P_{total} = \rho(A_d P_a + (1 - (A_d)))P_{sleep}. \tag{2.1}$$

Where, ρ is BS density, A_d is the ratio of active BSs density to that of all BSs and P_a is active BS power.

By modifying the hardware, software, or by using the BS sleep mode, the energy consumption can be reduced. In Ref. [33], the access network employs cyclic sleep modes for energy conservation. Numerous methods for incorporating BS sleep mode have been surveyed in Ref. [34]. Additionally, BS deployment and planning[35] are important for energy-saving. A very popular energy-saving method in 4G is discontinuous transmission and reception by turning off transceivers in absence of data transmission or reception.[36] In Ref. [37], the authors have investigated on the network power consumption using various deployment strategies. The authors in Ref. [38] have proposed static sleep mode of BS without considering traffic and network parameter change. A resource-on-demand (RoD) strategy is suggested by the authors in Ref. [39], where a cluster head is in charge of making sure that a set of WLANs is covered and the cluster is kept off to save energy during low loads. Under varying traffic conditions. Ref. [18] proposes adjusting BS modes (sleep/working). The authors in Ref. [40] propose the task of lowering the BSs supply power. The authors describe the inverse water-filling algorithm, which uses resource allocation using power control, antenna adaptation, and sleep modes. Power savings of 25 to 40% are possible with the help of this algorithm. Cellular network EE can be increased by making use of coordinated multipoint transmission.[41] As suggested in Ref. [42], BSs performance can be improved through multicell coordination for Greener Networks.

The authors in Ref. [40] propose the task of lowering the BSs supply power. The authors describe the inverse water-filling algorithm, which uses resource allocation using power control, antenna adaptation, and sleep modes. Power savings of 25 to 40% are possible with the help of this algorithm. Cellular network EE can be increased by making use of coordinated multipoint transmission.[41] As suggested in Ref. [42], BSs performance can be improved through multicell coordination for Greener Networks. Reduced BS density within cellular networks is another option for reducing BS power consumption. Such occurrences, however, can lead to coverage decadence. In response to the sharp increase in customers and the traffic rate,[43] the cell zooming technique for GREEN cellular networks has evolved. Cell zooming allows for adaptive cell size adjustment in response to traffic variations, reducing network energy consumption. A Cell Zooming situation is portrayed in Figure 2.5(a), in the group of five cells. The five cells in Figure 2.5(a), represent a cell zooming situation.

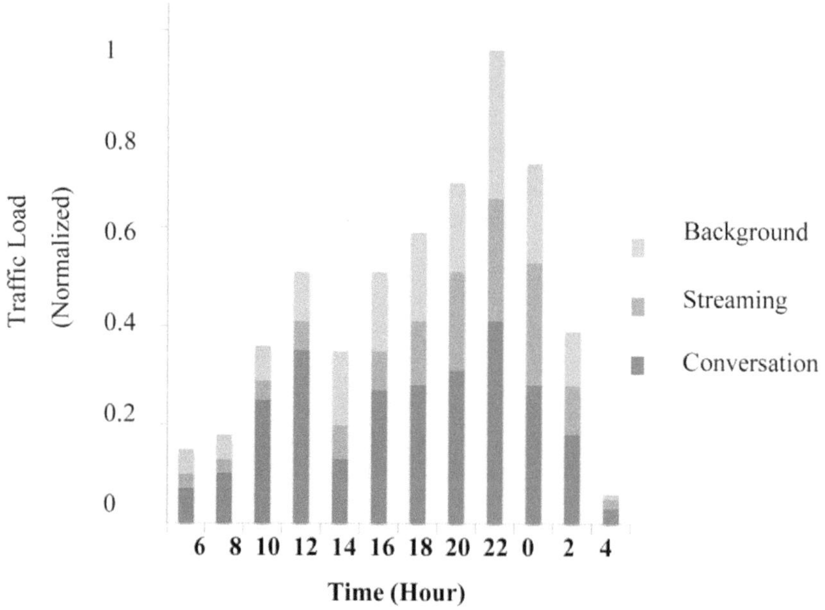

FIGURE 2.4 Traffic load profile of a day.

Every cell has a BS in the middle, and users are dispersed at random. Congestion develops when mobile users migrate toward the central cell. The cell shrinks inward to escape this crowded state (Figure 2.5(b)). Another scenario is that the users migrate out into the next cells, causing congestion in them. In order to guarantee coverage for every user, the central cell can be zoomed out while the surrounding ones are zoomed in (Figure 2.5(c)) to handle the users. In fact, the cell in the center may go to sleep mode to ensure less energy usage (Figure 2.5(d)).

In order to ensure this, the remaining cells either zoom out or work together to ensure coverage. The cell zooming server (CS) manages cell zooming.[44]

2.4 TECHNIQUES FOR POWER ALLOCATION AND SOME ENERGY-EFFICIENT (GREEN) METRICS

Different power allocation techniques are used to increase the EE of mobile networks and thereby transition to GREEN Network. Two of the most popular power allocation algorithms, water filling (WF) and equal power allocation (EPA). In order to establish an energy-efficient network architecture, the crucial performance measures must also be properly identified. A long- term research and development project requires the examination of many metrics since they enable performance comparisons across various methods and ideas. Here, some key performance indicators for assessing the EE of future networks have been presented.

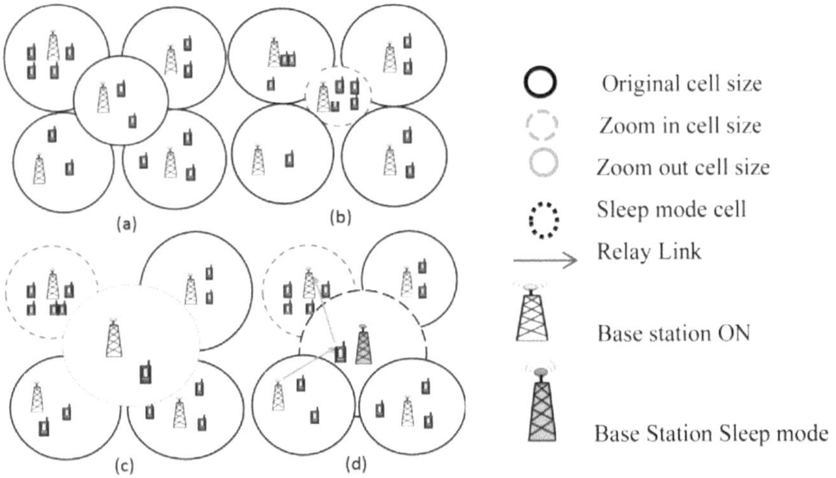

FIGURE 2.5 (a) Cell layout; (b) cell zooms in; (c) cell zooms out; (d) sleep mode with information transfer through relaying.

2.4.1 STRATEGIES FOR POWER ALLOCATION

With adequate power allocation, cellular network efficiency can be increased.

1. *Equal Power Allocation or EPA Scheme:* When a user *i* with total power P_{total} is required to transmit using the EPA technique with a certain number of resource blocks *Ni*, its transmission power *Pi* is determined by

$$Pi = P_{total}/Ni \tag{2.2}$$

2. *Water filling Algorithm:* This iterative power allocation scheme is an algorithm that quickly converges. This power allocation approach is applied to ISI channels, and channels that employ frequency

selective fading in order to maximize capacity.[45] The highest data rate that may be achieved, subject to the input power restriction. Graph theory, game theory, and the branch and bound algorithm[46] are further methods for power allocation and optimization. These have less overhead in terms of calculation.

2.4.2 THE GREEN METRICS

An effective solution requires a detailed analysis of the network performance and the energy savings. In this case, energy measurements are crucial. These aid in figuring out how much power is consumed overall by the network. It is possible to categorize EE metrics at the facility, equipment, and network levels. EE metrics are discussed in Ref. [47]. Each level measurement is specified in a different method. A handful of GREEN metrics are discussed here.

1. *Energy Efficiency (EE):* It is the ratio of overall data rate to the total power consumption, taking into account the power used by the BS.
2. *Area Energy Efficiency (AEE):* When a cellular heterogeneous network with a dense deployment of small cells (micro, Pico, and Femto) is taken into consideration, area energy efficiency is a critical performance measure. It takes into account how the size of cells affects energy effectiveness. It is the ratio of EE and cell area.
3. *Outage Probability:* It is defined as the probability of a user for which the downlink signal to interference noise ratio is less than the target signal to interference noise ratio.
4. *Spectral Efficiency:* It is the highest data rate that can be consistently sent through the channel with zero error.
5. *Energy Harvest Ratio:* It is described as the length of time spent harvesting energy during a specified period of time.
6. *Average Sum Rate (ASR):* It is obtained from the transmission rates of all users within the network.

A single indicator is insufficient to meet the requirement for GREEN measuring in a cellular network. When combined, all of these things can be useful. The next section comprises of different technologies in GREEN Communication.

2.5 ENERGY-EFFICIENT TECHNOLOGIES IN 5G

In the next decade, the amount of devices linked via the cloud, that can access data anywhere and at any time is to cross 50 billion. This growth is having a negative impact on the environment. However, such an increase is unsustainable from the mobile network operators' (MNOs') economic standpoint. Various strategies, such as the WF algorithm, Game theory, EPA algorithms,[48–50] etc. have been studied to reduce network power usage. The 5G networks, as well as subsequent networks, are facilitating network upgrading. Such networks allow for a constant total power usage. This goal is achievable by reducing energy consumption that is not directly connected to the transmission of information. Communication and information technologies by 2030 have the potential to reduce CO_2 emissions by 20%. Figure 2.6, depicts ICT-enabled advances that are predicted to reduce CO_2 emissions by 2030 in several sectors.

Various 5G technologies that enable energy conservation are shown in Figure 2.7. This section goes into great length about the different technologies.

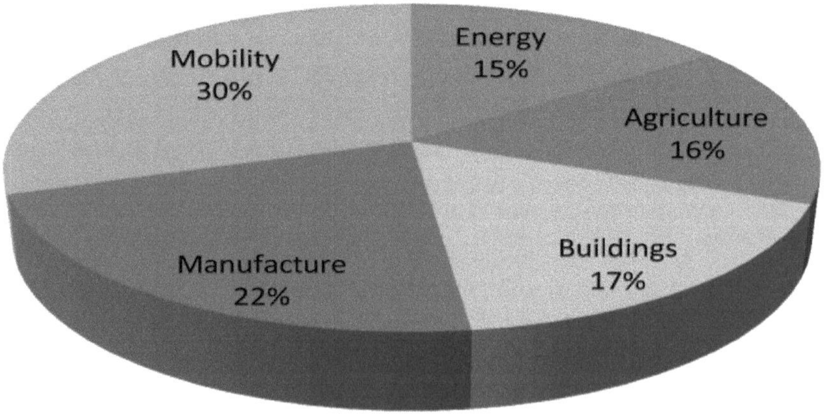

FIGURE 2.6 By 2030, carbon footprint reduction through smart technologies.

2.5.1 DEVICE-TO-DEVICE (D2D) COMMUNICATION

For 5G networks, D2D (Figure 2.8) is a useful technology.[51] It improves the connection reliability between users by establishing direct links between them and thereby reducing latency.

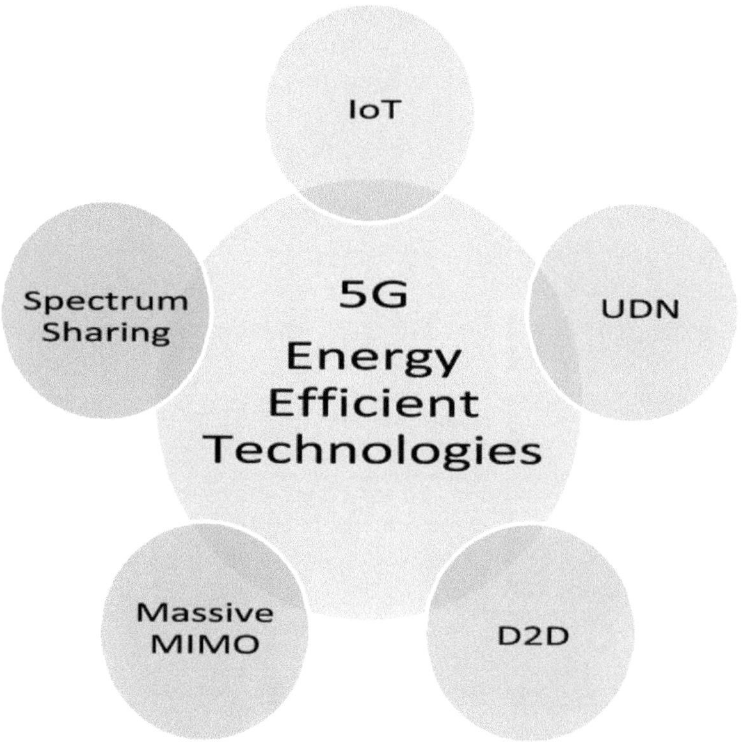

FIGURE 2.7 Energy-efficient technologies.

A variety of D2D communication-related problems have been solved by Ref. [52]. Transferring traffic to direct links relieves the BS's workload and encourages BS sleep. An approach like this aids in BS and user equipment using less energy. Power control is crucial in mobile networks because it boosts system capacity and reduces interference.[53] Power control is covered in Ref. [54] along with how D2D communication might make it better. Many methods for effective power management have been put out in the literature. A variety of measures are used to assess power control in cellular networks, with EE being the most used one. Networks that are wired and wireless are both undergoing energy-saving initiatives. The primary energy users in wireless networks are the access networks, which comprise their main parts and use more than 70% of all available energy.[55] When users are near cell edges in a communication system, the BS must emit incredibly high levels of power. There is a chance that power levels will exceed their upper thresholds, which would be completely improper for cellular networks. To save on system power, D2D communication can be used as a cellular network overlay. It is a practical way to sufficiently reduce transmission power.

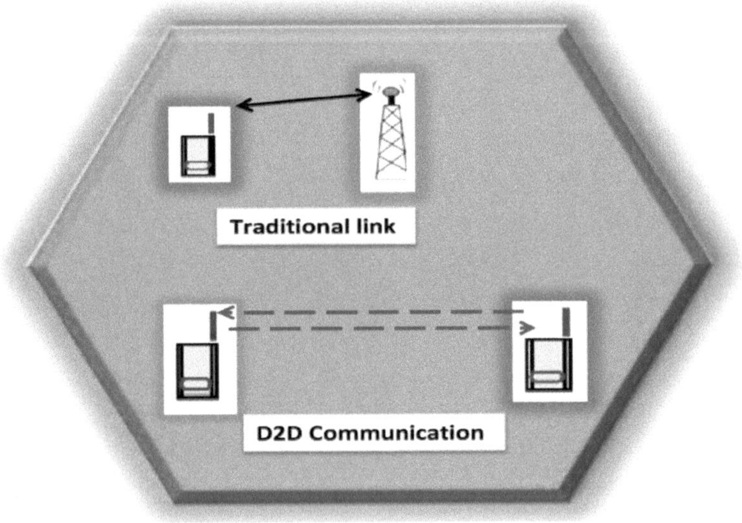

FIGURE 2.8 D2D communication.

Three operating modes are available to D2D users: cellular, dedicated, and reuse. For optimal system performance, selecting the appropriate D2D mode is essential. During the selection of modes, EE must be given top priority. In Ref. [56], the authors examine energy-efficient mode switching and address the EE optimization issue in each of the three modes. The suggested reuse mode technique plays a crucial part in maximizing EE for D2D users. Additionally, effective interference management is required for the coexistence of cellular and D2D lines. In Ref. [57], a joint mode selection and power allocation approach is put forth. It takes both system capacity and transmission power into account to maximize the utility function. The determination of suboptimal power levels is done at the beginning followed by the selection of mode. The mode with the highest level of power efficiency is chosen. This plan has the potential to enable.

D2D energy-efficient communication. The coexistence of cellular and D2D communication can be maintained by decreasing interference and adjusting power levels. By integrating power and data rate optimization, D2D users' EE is boosted in Ref. [58] when cellular users are present. The sub-gradient method and the Lagrangian duality theory are both employed to determine the optimal solution for EE maximization. With this initiative, users of cellular and D2D communication will be treated fairly in proportion.

The focus of the authors of Ref. [59] is on using a derivative-based strategy to maximize EE over the whole network. In order to calculate EE, stochastic geometry theories are required. These theories assume that the distribution of users throughout the different bands is random. A nonconvex optimization algorithm used is highly helpful and has minimal computing complexity. Additionally,[60] uses iteratively combined resource allocation and power control to maximize EE. It is suggested to use a two-layer strategy in which the first layer power values are initially derived via a series of maximization problems. In the next layer, an optimal solution for the specified problem is then computed. The method is centralized and has a high administrative cost. According to Ref. [61], there are three different ways for two users to communicate: conventional cellular, multihop D2D or direct D2D connection. For the three modes, average spectral and EE is examined. The ideal power level of every individual subscriber equipment is also considered for each mode. Based on stochastic geometry, distributed and centralized resource allocation methods in Ref. [62] have been suggested. The primary variables for determining the network performance are evaluated with regard to a D2D communication that underlies cellular networks. Resources can be shared between D2D and cellular users in three different ways: orthogonal, non-orthogonal, and cellular. Methods for power distribution for these three modes are provided in Ref. [63] to improve EE when operating under the maximum power limit. Resource distribution is controlled by the BS. For the non-orthogonal mode, noncooperative game theory is used, while for the other two modes, quasi-concave function optimization is used.

In order to address a nonconvex resource allocation problem for EE, a heterogeneous network with integrated D2D and cellular communication is taken into consideration in Ref. [64]. Fractional programming is used to convert a nonconvex issue to a convex one. The Coordinate Ascent Method and the Dinkelbach Method both contribute to further solutions.

The simulation results depicted that there was a greater EE when cellular networks and D2D communication coexisted. The EE maximization problem is formulated by the authors in Ref. [65] in the context of resource allocation and cell selection for heterogeneous networks (HetNets). The proposed method uses a linearly converging algorithm.

The authors in Ref. [66] make a suggestion for an energy-efficient power control technique for D2D communication that supports mobile networks. When the EE and minimum data rate are the objectives, the network's QoS requirements are guaranteed. Generalized and nonconcave fractional programming issues are used to develop both individual and total EE, respectively. The centralization of this design lowers its complexity and associated computational requirements. It is demonstrated that EE performance for superior and poorer systems performs pretty similarly. The authors of Ref. [67] offer an effective power optimization strategy for a considerable reduction in overall BS power consumption. To integrate D2D communication, an orthogonal frequency-division multiple access (multiuser version of OFDM) technique is suggested for subcarrier, bit allocation, as well as correct mode selection. As previously indicated, proper power regulation can help to control interference between D2D and cellular users. The power of downlink is carefully adjusted for networks having a large number of D2D pairs.

According to Ref. [68], greedy sum rate maximization is carried out when the maximum transmit power is constrained. Cellular users are given top priority under the maximum power restriction and are also guaranteed a minimum transmission rate. The transmission rate is therefore limited to the highest achievable for the system using the highest modulation and coding technique. In order to maximize EE, a multiobjective optimization problem (MOOP) was formulated[69] and used to examine a power control architecture[70] that discusses how to allocate resources and choose modes that are EE. The sum rate maximization problem, according to the authors of Ref. [71], is the most effective technique to match cellular and D2D users and also deal with power allocation to the D2D links.

To extend the lifespan of the devices, EH[72,73] is necessary. In Ref. [74], EH and cognitive-based D2D communication is investigated. The energy harvested by D2D transmission plays a role in spectrum sensing. Later on, when D2D transmitters have sufficient energy and harvesting is done, the spectrum detection by them takes into account two spectrum access schemes (random and prioritized spectrum access). Stochastic geometry is used to evaluate the proposed system. Due to the EH circuitry in each D2D transmitter, radio frequency power from both the up-link and down-link channels is efficiently captured. In the company of cognitive D2D communication, this solution achieves an acceptable level of QoS.

Relay-based transmission has been suggested by 3GPP[75] as a way to improve network quality. There may be fixed (Type I) or portable relays employed (Type II). For the relay to execute the transmission, additional power is required. In this situation, the role of EH is vital. Energy harvested power profile has some randomness attached to it, which should be taken into account in these networks. In Ref. [76], considers a heterogeneous network with D2D communication support and EH capabilities. Relays are thought to have EH capabilities and can transmit data to D2D users using the collected energy. Each user equipment EH is modeled using a Markov chain. The use of harvested energy can boost network performance of D2D communication. The proposed method can save a significant amount of energy. D2D, BS sleeping mode, and the other technologies covered in the next sections can all effectively support the GREEN Communication scheme.

2.5.2 ULTRA-DENSE NETWORKS

New technologies are being developed to provide abundant wireless access. The emergence of ultra dense networks (UDNs) ensures seamless communication. UDNs entail the installation of numerous small cells in heavily populated areas. Small cells are wireless access points with low power and low cost, which increase capacity and coverage while off-loading data.[77] The deployment of micro, Pico, and Femto cells plays a role in indoor coverage improvement. The spectrum efficiency and EE of cellular networks have significantly increased with the small cells for coverage expansion. Figure 2.9, displays the various cell sizes and their related transmission powers. In Ref. [78], a distributed architecture for 5G UDNs was put forth. A thorough analysis of UDNs has been provided in Ref. [79]. The literature has reviewed various potential systems for energy- efficient UDNs. Femto cells are a crucial component in meeting the cellular networks' high data rate requirements. Such cells consume too much energy when they are deployed densely, necessitating an increase in network EE. In order to accomplish the goal, a clustering strategy is suggested in Ref. [80]. The semi-centralized framework is developed with straightforward algorithms. The suggested scheme has low levels of complexity and excellent computing efficiency. The use of stochastic geometry to maximize EE in UDNs has been proposed in Ref. [81]. In order to increase network capacity and coverage, the Pico cells have emerged as an excellent low-cost and low-power alternative. However, the intensive application of Pico cell BSs results in excessive consumption of energy. In Ref. [82], the authors examine both EE and area EE while taking the power outputs of macro and Pico BSs into consideration.

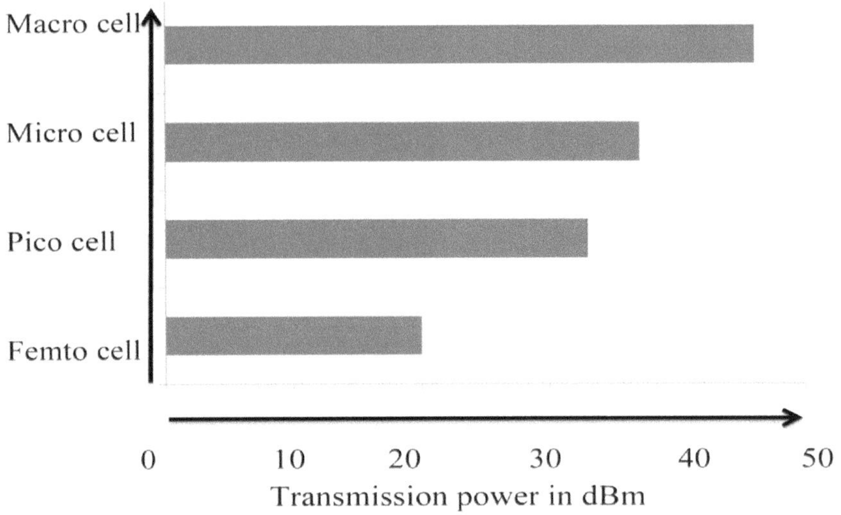

FIGURE 2.9 Transmission power in different cell size.

According to the current scenario, the implementation of UDNs will enable network intensification to reach unrestricted levels. The network's EE is analyzed by the authors in Ref. [83] for various outdoor and indoor deployment situations. Extreme intensification levels have been examined. In comparison to outdoor situations, the simulation findings demonstrate great EE of densely deployed indoor scenarios. In terms of cost and EE, such a strategy is beneficial for real-world systems. Spectrum, energy, and cost efficiencies are going to be crucial performance factors for the next-generation of GREEN Ultra Dense Networks. In Ref. [84], all of these metrics are mathematically represented into a comprehensive and single model. All three efficiencies are taken into account while designing a weighted utility function, followed by a Nash product utility function. The suggested design helps in the achievement of ideal equilibrium solutions. In Ref. [85], a method for enhancing efficiency of the spectrum and EE that supports the sleep mode of BS during low traffic period is examined. Spectrum share between macro cells and small cells results interference. In Ref. [86], a utility function for spectrum and EE was developed to reduce interference and enhances energy conservation.

A unique method for combining power control and user scheduling has been suggested in Ref. [87] for optimizing EE in UDNs. A dynamic stochastic game use guarantees QoS to users within the network.

Frequency reuse factor is a crucial aspect when cellular network dimensions are being determined. The majority of the existing research on BS deployment takes into account frequency reuse, which degrades spectrum and EE owing to inter-cell interference. In order to maximize the frequency reuse factor, a random and dense deployment of BS is taken into consideration.[88] Cell breathing bridges the distance between the quantity of resource availability as well as the demand. A high resolution cell breathing (HiRCB) technique has been developed in Ref. [89] to increase the energy effectiveness of UDNs. In this work, the traffic differences in both space and time have been taken into consideration. With the suggested technique, EE is increased by at least 50%. EH from RF sources is a promising approach under evaluation for wireless device charging. Ambient EH and specialized EH are two types of EH.[90] discusses recent accomplishments in this discipline as well as the associated flaws, and thereby assisting researchers in identifying research gaps. The authors assessed the availability of EH and suggested method for optimizing network settings.[91] proposes a precoding approach for improving the EE of small cell networks.

Massive MIMO usage further improves network performance. With a BS density of 100 per km^2, numerous algorithms listed in Refs. [81–85] validate EE improvement.

2.5.3 *MASSIVE MIMO*

The 4G networks have benefited greatly from multiple input multiple-output (MIMO) technology. For several gigabytes transmission per second massive MIMO emerges as a solution.[92–94] The advantages of massive MIMO technology are numerous. A pictorial representation of the same can be seen in Figure 2.10. It offers all of the advantages of MIMO, but on a greater scale. Numerous benefits of massive MIMO exist, such as higher throughput, improved spectrum efficiency, robustness against jamming, decreased latency, enormous capacity increases, and improved EE.[94] Massive MIMO utilizes less energy than MIMO.[95] With Massive MIMO, the amount of radiation can be reduced up to 1000 times.[96] In Ref. [97], a multicell massive MIMO system is taken into consideration with the intention of lowering overall power usage. Authors have taken into account the user equipment's (UEs) parameters and the BS power constraints to minimize the transmission power. Work in Ref. [98] analyses yet another multicell massive MIMO network. Maximum Ration Transmission and zero-forcing beam forming are used to examine the scaling laws of EE. Further, the study of joint pilot assignment and resource allocation for maximization of system EE is presented in Ref. [99]. For power optimization, pilot contamination is specifically taken into account. A sequential convex optimization strategy is used for convex optimization problems. With regard to system EE and total rate, the proposed approach can produce considerably superior results. Researchers in Ref. [100], makes a suggestion for a distributed power control algorithm. This work considered a scenario with multiple users and multiple cells, where each user is able to optimize their EE. Therefore, computational complexity was significantly lowered. The evolutionary game theory (EGT) in Ref. [101] enhances fairness among users. To balance between bit error rate and spectrum efficiency, antenna grouping is necessary. In Ref. [102],

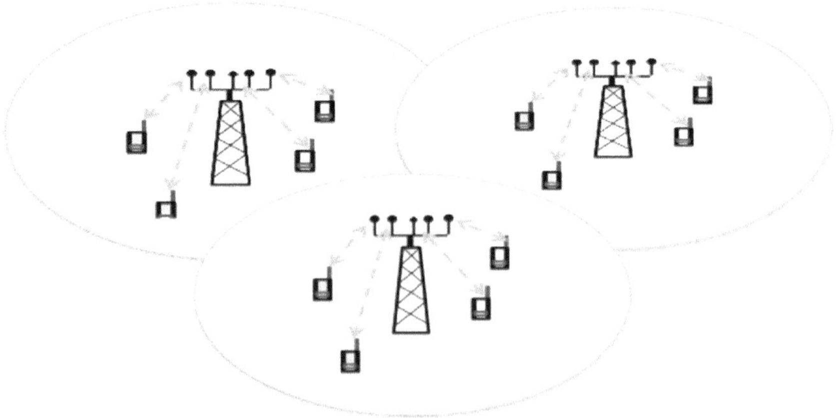

FIGURE 2.10 Massive MIMO

the beam form by antenna grouping is proposed. In a MIMO system, the work done in Ref. [103] enhances the EE by utilizing spatial multiplexing and beam forming due to selected antenna grouping. Hence, at the BSs, numerous antennas could increase complexity issues. These still need to be studied by the researchers. Antenna muting is another method for using antennas to save energy.[104,105]

2.5.4 SPECTRUM SHARING

The operation of cellular network required dedicated licensed spectrum. To mitigate the demand of exponentially increased users many technologies have emerged. Therefore, in today's wireless communication, both unlicensed and shared licensed spectrum are available to support high data transmission. Moreover, as the radio spectrum is overcrowded, dynamic spectrum allocation is required. As deployments get more dense, the regulatory solutions for 5G and beyond is a need to prevent low spectrum utilization efficiency. These various sharing mechanisms (underlay/overlay/mixed), enables the unlicensed secondary users to access the underused portions the licensed spectrum. It is a useful technology for managing spectral efficiency and maximizing EE simultaneously.[106] The advanced technologies for spectrum sharing and operation modes have been covered in Ref. [107]. Spectrum-sharing methods can be roughly divided into distributed and centralized categories.[107] According to the customers' regulatory requirements, hardware-based spectrum-sharing systems have been researched in Ref. [108]. It is a practical way to deal with the issue of escalating consumer demands.[109] discusses the evolution of several contemporary circuit optimization techniques while taking power economy and spectrum performance into account. These systems aim to attain high capacity through flexible spectrum utilization. In this literature, power level optimization in spectrum-sharing networks has been explored. Stochastic geometry is used in Ref. [110] to study cooperative spectrum sharing. The best cooperation mechanism, which the authors of Ref. [111] suggest, is capable of both EH and data transmission within a timeslot in a network. In this system, both types of users have the ability to gather energy by harvesting process. The ability of the secondary users to serve as relays during the primary user data transmission improves throughput and speeds up the primary users' data transmission rate. In

Ref. [112], as secondary users serve as relays, prime users' transmission is finished much earlier than expected. Cat swarm optimization (CSO) is used to achieve the same. Spectral efficiency and EE have been examined for a two-tier heterogeneous network (HetNet) in Ref. [113] under a spectrum-sharing scenario. The proposed design places a high priority on the requirement for a workable method for multitier cellular networks. An effective operational regime is chosen using the Karush-Kuhn-Tucker (KKT) requirements. According to analysis, larger spectrum efficiency and EE are produced by Femto-tier BSs with higher density; however, this effect diminishes as BS power consumption and load increase. Some techniques have been discussion for OFDM-based spectrum sharing in Refs. [114, 115]. In spectrum-sharing networks.[116] with multicarrier, the downlink radio resource allocation (RRA) problem was studied using the idea of time-averaging window. In addition to it, Partial spectrum reuse is a useful method for increasing both spectrum and EE.[117] In heterogeneous networks (HetNet), it is successful in preventing inter-cell interference. For the best power distribution, the authors of Ref. [118] look into a cross-interfering spectrum-sharing scheme. In order to implement the best power control measures, double-threshold water filling (DT-WF) and double-threshold constant water filling (DTCP-WF) have been developed. Therefore, spectrum sharing is anticipated to revolutionize next-generation technologies and make a significant contribution to energy conservation.

2.5.5 INTERNET OF THINGS (IOT)

With the rise of IoT, billions of people in less time will be connected. The usage of RFIDs and data transfer for linking anything in the world is part of computerized interconnectivity. Wireless sensor networks (WSNs) are an important IoT platform that may support a wide range of applications and fulfill the vision of a smarter world. IoT provides a competent option for achieving the goal of a minimum costing of carbon and saving energy as well. In an IoT infrastructure, the optimal power management at the sensor nodes is essential. In each sensor node, EE is an important parameter for proper functioning.

As these sensor nodes work on low power therefore energy-saving technologies are a dire need. The work in Ref. [119] proposes a method of faster data gathering from the distant nodes and slower from its adjacent.

The approach achieves an increase in EE of up to 19%. In Refs. [119–121], another strategy for increased energy economy was used, in which not so relevant and superfluous sensors in the network were switched to mode-sleep. Cellular partition zooming (CPZ) and pre-caching mechanism also contributed to EE improvement of 5G It is a flexible solution and can be used in both wired and wireless part of a networks. Some more research work such as Ref. [122] proposes the particle swarm optimization (PSO) technique for saving energy in sensor networks. With the suggested method, a power savings of almost 1 dBm is feasible. It has been tried in a variety of circumstances, although it is not convergent in all of them, necessitating additional exploration. Ref. [123] discusses a power management technique based on Vague sets that allows adaptive power control. Vague sets are a fuzzy theory that is utilized for fuzzy information. Vague set theory technique results in minimal energy consumption while ensuring reliable and efficient communication. To ensure better QoS and throughput, D2D communication is one among many technologies being utilized in combination with the IoT.[124] In Ref. [125] for D2D communication, a centralized method with the distributed scheme for channel selection and optimal power allocation is used to avoid interference. The inclusion of game theory in this concept results in a significant increase in EE. To reduce energy consumption in industries, a work in Ref. [126] is an amalgamation of cognitive radio (CR) with the industrial Internet of things (IIoT). An adaptive power control method employing nonlinear programming is proposed for better results. The devices in IoT environment can function as energy harvesting sources. Numerous methods of code upgrade for programmable devices can be seen in Ref. [127], which can significantly improve EE and drop the resource demand. In point of fact IoT is aiming toward smart energy management which is evident in Refs. [128, 129]. The EE of cellular networks has been improved by the implementation of several measures. A distributed topology control approach has been put forth in Ref. [130]. GREEN Cloud RAN guarantees improved EE in UDN and thereby helps in reducing energy consumption bills. This topic has been covered from an implementation standpoint in Ref. [131]. In Ref. [132], a test bed for GREEN cities has been examined. In this context, it may be concluded that the application of IoT has the potential for significant energy savings. The idea of energy Internet (EI) has been suggested to achieve the balance between supply and consumption. Although this is helpful for NGNs, there are still some issues that require specific attention.[133] There has been discussion of the many key technologies for 5G networks.

Aside from these technologies, millimeter wave (mmWave) communication is another growing technology. It can accurately assess beam forming quality and increase system EE.[134] Massive MIMO and mmWave communication may considerably enhance the BS coverage area, improving network QoS. The implementation of spectrum sharing can be considerably improved with a large number of antennas.[135] Different strategies presented have resulted in considerable reductions in MNO expenses and CO_2 emissions.[136] It is obvious that all of the developing technologies are interconnected and helpful for GREEN Networking. However, before they are put into practice, a few open problems need to be seriously addressed. GREEN communication necessitates green coordination among network providers. This is an excellent opportunity to get hands-on experience with the approaches.

2.6 RESEARCH CHALLENGES

The several energy-saving methods that have already been discussed in this work are beneficial for 5G. The paper discusses a variety of approaches, each of which is also thoroughly examined. Combining all of them into one comprehensive approach has the potential to transform GREEN Communication. The needless energy loss that occurs when the BS is turned on and off must be taken into consideration. When using BS sleep modes, the awakening time must be configured. This requires a different period for the BS to activate as the load increases. The growth of the IoT will result in a significant increase in machine-type communication (MTC). This will result in network congestion. A congested network will result in a wastage of power, energy, and an increase in computational complexity. Design on congestion Controller[138–142] for smart power management can be research direction.

The use of BS on/off mechanisms can be daunting in 5G networks. Users of the switched-off cell are also assisted during sleep intervals by the BS of the subsequent cell. In order to support its users, the operator whose BSs are disabled must pay the adjacent operator. If the pricing is not economical for the operators, there is no use in employing sleep modes to save energy. The emerging networks have a random aspect as well. THE stochastic geometry and random matrix theory approaches are thereby helpful for the study of random behavior. Researchers in this subject will have a free field to explore GREEN Networks by letting the devices draw on their existing knowledge and react suitably through self-organization. The proliferation of technology also increases harmful radiation and carbon footprint levels. Novel antenna configurations and less radiate surface materials can actively reduce the harmful radiation from user equipment. Since a significant quantity of CO_2 is released during the production process, recyclable materials can help in reducing the carbon footprint. One of the main deployment concerns for UDNs is powering and backhauling. Here, the power requirements of BS in every small cell must be met. When examining this problem, the idea of self-backhauling has emerged. This is still an unexplored area of study. Small cells that are fueled by renewable energy sources are becoming widely used. However, BS sleep modes have not yet been supported by such deployments. End-to-end latency is an important measure that needs to be handled.

An extremely energy-efficient method is massive MIMO. However, the great number of antennas results in a significant amount of overhead, which consumes power inefficiently. Another major drawback is the increased price. Massive MIMO involves intensive processing, that calls for further thought. The work in Ref. [137] is an energy-efficient power allocation technique under spectrum-sharing scenario. Spectrum sharing can, as already indicated, significantly improve EE. However, while taking security into account, regulatory restrictions on spectrum access need to be addressed. Economic issues with spectrum sharing are also quite significant.

As energy cannot be protected, it is extremely vulnerable to security risks. Energy forgery attacks occur when the energy status of devices in a network is altered. Therefore, concerns about privacy in relation to energy states must be addressed. Additionally, backups of information and data can be made, but not of energy. The researchers are concerned about this topic. An in-depth study has been done on EH, which involves drawing energy from RF environmental sources. Renewable energy resources are dynamic in nature, for instance, different times of day and night can yield different results when harvesting solar energy. In addition, it is important to efficiently distribute the energy to the users. As a result, there are still several scientific barriers in the way of a totally GREEN Cellular Network.

2.7 CONCLUSION

The mobile network EE is an edgy for the network operators. A rise in customers and gadgets as a result of technological progress results in an increase in cost but also has negative effects on the environment and health, including rising CO_2 levels. In the study, these elements were initially briefly discussed. Several techniques for controlling the power inside the BS have been investigated because it consumes the majority of the energy. Massive MIMO, device-to-device (D2D) connectivity, spectrum sharing, and the IoT are technologies supported by the upcoming 5G networks. Each of these has undergone extensive analysis from this perspective, and they can all facilitate GREEN communication in cellular networks in the future. Despite having benefits of GREEN Networking, there are several problems that have been explored. This opens the path for additional study in this area. This study makes a suggestion for employing spectrum sharing to extend battery life. It is an excellent strategy for preserving user battery life in cellular networks without affecting QoS. As previously said, network security is one feature of all growing technologies that cannot be compromised. Secure power optimization is critical for GREEN cellular communication. For NGNs, this is still an open study topic.

KEYWORDS

- **Internet of things (IoT)**
- **D2D communication**
- **energy efficiency (EE)**
- **massive MIMO**
- **ultra-dense networks (UDNs)**

REFERENCES

1. Monserrat, J. F.; Mange, G.; Braun, V.; Tullberg, H.; Zimmermann, G.; Bulakci, O. METIS Research Advances Towards the 5G Mobile and Wireless System Definition. *EURASIP J. Wireless Commun. Netw.* **2015,** *53* (1), 20.
2. NGMN. 5G White Paper, 2015.

3. Perera, C.; Liu, C. H.; Jayawardena, S.; Chen, M. A Survey on Internet of Things from Industrial Market Perspective. *IEEE Access* **2014,** *2*, 1660–**1679**.
4. Li, Q. C.; Niu, H.; Papathanassiou, A.; Wu, G. 5G Network Capacity: Key Elements and Technologies. *IEEE Veh. Technol. Mag.* **2014,** *9*, 71–78.
5. Gavrilovska, L.; Rakovic, V.; Atanasovski, V. Visions Towards 5G: Technical Requirements and Potential Enablers. *Wireless Pers. Commun.* **2016,** *87*, 731–757.
6. Soldani, D.; Manzalini, A. Horizon 2020 and Beyond: On the 5G Operating System for a True Digital Society. *IEEE Veh. Technol. Mag.* **2015,** *10* (1), 32–42.
7. Thakare, P.; Ambudkar, B. Study and Analysis of CO_2 Emission Control Methods for Wireless Networks. In: *2nd International Conference on Communication and Electronics Systems (ICCES)*; **2017**; pp 742–745.
8. Kumar Jadav, N.; Gupta, R.; Tanwar, S. A Survey on Energy-Efficient Resource Allocation Schemes in Device-to-Device Communication. *Int. J. Commun. Syst.* **2022,** *35* (8), e**5112**.
9. Luo, F. L.; Zhang, C. *Energy-Efficient Resource Allocation in 5G with Application to D2D*, 2016; 456–482.
10. Abidrabbu, S. S.; Arslan, H. Energy-Efficient Resource Allocation for 5G Cognitive Radio Noma Using Game Theory. In: *2021 IEEE Wireless Communications and Networking Conference (WCNC)*, 2021; pp 1–5.
11. Son, K.; Kim, H.; Yi, Y.; Krishnamachari, B. Base Station Operation and User Association Mechanisms for Energy-Delay Tradeoffs in Green Cellular Networks. *IEEE J. Select. Areas Commun.* **2011,** *29* (8), 1525–**1536**.
12. Marsan, M. A.; Bucalo, G.; Di Caro, A.; Meo, M.; Zhang, Y. Towards Zero Grid Electricity Networking: Powering BSS with Renewable Energy Sources. In: *2013 IEEE International Conference on Communications Workshops (ICC)*, 2016; pp 596–601.
13. Han, T.; Ansari, N. Powering Mobile Networks with Green Energy. *IEEE Wireless Commun.* **2014,** *21* (1), 90–96.
14. Ng, D. W. K.; Lo, E. S.; Schober, R. Energy-Efficient Resource Allocation in OFDMA Systems with Hybrid Energy Harvesting Base Station. *IEEE Trans. Wireless Commun.* **2013,** *12* (7), 3412–**3427**.
15. Lu, X.; Wang, P.; Niyato, D.; Kim, D. I.; Han, Z. Wireless Networks with RF Energy Harvesting: A Contemporary Survey. *IEEE Commun. Surv. Tutor.* **2015,** *17* (2), 757–789.
16. Bhargava, V. K.; Leon-Garcia, A. Green Cellular Networks: A Survey, Some Research Issues and Challenges. In: *2012 26th Biennial Symposium on Communications (QBSC)*, 2012; pp 1–2.
17. Abrol, A.; Jha, R. K. Power Optimization in 5G Networks: A Step Towards Green Communication. *IEEE Access* **2016,** *4*, 1355–**1374**.
18. Garroppo, R. G.; ScutellÃă, M. G.; D'Andreagiovanni, F. Robust Green Wireless Local Area Networks: A Matheuristic Approach. *J. Netw. Comput. App,* **2020,** *163*, 10**2657**. https://www.sciencedirect.com/science/article/pii/S1084804520301314
19. Buzzi, S.; L, C.-L.; Klein, T. E.; Poor, H. V.; Yang, C.; Zappone, A. A Survey of Energy-Efficient Techniques for 5G Networks and Challenges Ahead. *IEEE J. Select. Areas Commun.* **2016,** *34* (4), 697–709.

20. Gupta, A.; Jha, R. K. A Survey of 5G Network: Architecture and Emerging Technologies. *IEEE Access* **2015,** *3*, 1206–**1232**.
21. Feng, D.; Jiang, C.; Lim, G.; Cimini, L. J.; Feng, G.; Li, G. Y. A Survey of Energy-Efficient Wireless Communications. *IEEE Commun. Surv. Tutor.* **2013,** *15* (1), 167–178.
22. Bolla, R.; Bruschi, R.; Davoli, F.; Cucchietti, F. Energy Efficiency in the Future Internet: A Survey of Existing Approaches and Trends in Energy-Aware Fixed Network Infrastructures. *IEEE Commun. Surv. Tutor.* **2011,** *13* (2), 223–244.
23. Fehske, A.; Fettweis, G.; Malmodin, J.; Biczok, G. The Global Footprint of Mobile Communications: The Ecological and Economic Perspective. *IEEE Commun. Mag.* **2011,** *49* (8), 55–62.
24. Mancuso, V.; Alouf, S. Reducing Costs and Pollution in Cellular Networks. *IEEE Commun. Mag.* **2011,** *49* (8), 63–71.
25. Correia, L. M.; Zeller, D.; Blume, O.; Ferling, D.; Jading, Y.; GÃs¸dor, I.; Auer, G.; Der Perre, L. V. Challenges and Enabling Technologies for Energy Aware Mobile Radio Networks. *IEEE Commun. Mag.* **2010,** *48* (11), 66–72.
26. Lopez-Perez, D.; De Domenico, A.; Piovesan, N.; Bao, H.; Xinli, G.; Qitao, S.; Debbah, M. A Survey on 5G Radio Access Network Energy Efficiency: Massive Mimo, Lean Carrier Design, Sleep Modes, and Machine Learning, 2021. https://arxiv.org/abs/2101.11246
27. Niu, Z. Tango: Traffic-Aware Network Planning and Green Operation. *IEEE Wireless Commun.* **2011,** *18* (5), 25–29.
28. Willkomm, D.; Machiraju, S.; Bolot, J.; Wolisz, A. Primary User Behavior in Cellular Networks and Implications for Dynamic Spectrum Access. *IEEE Commun. Mag.* **2009,** *47* (3), 88–95.
29. Xiao, Z.; Fu, X.; Zhang, L.; Goh, R. S. M. Traffic Pattern Mining and Forecasting Technologies in Maritime Traffic Service Networks: A Comprehensive Survey. *IEEE Trans. Intell. Transport. Syst.* **2020,** *21* (5), 1796–**1825**.
30. Williams, L.; Sovacool, B. K.; Foxon, T. J. The Energy Use Implications of 5G: Reviewing Whole Network Operational Energy, Embodied Energy, and Indirect Effects. *Renew. Sustain. Energy Rev.* **2022,** *157*, 11**2033**. https://www.sciencedirect.com/science/article/pii/S1364032121012958
31. Soh, Y. S.; Quek, T. Q. S.; Kountouris, M.; Shin, H. Energy Efficient Heterogeneous Cellular Networks. *IEEE J. Select. Areas Commun.* **2013,** *31* (5), 840–850.
32. Alsharif, M. H.; Kim, J.; Kim, J. H. Green and Sustainable Cellular Base Stations: An Overview and Future Research Directions. *Energies* **2017,** *10* (5). https://www.mdpi.com/1996–1073/10/5/587
33. Anthapadmanabhan, N. P.; Dinh, N.; Walid, A.; van Wijngaarden, A. J. Analysis of a Probing-Based Cyclic Sleep Mechanism for Passive Optical Networks. In: *2013 IEEE Global Communications Conference (GLOBECOM)*, 2013; pp 2543–**2548**.
34. Wu, J.; Zhang, Y.; Zukerman, M.; Yung, E. K.-N. Energy-Efficient Base-Stations Sleep-Mode Techniques in Green Cellular Networks: A Survey. *IEEE Commun. Surv. Tutor.* **2015,** *17* (2), 803–826.
35. Weicker, N.; Szabo, G.; Weicker, K.; Widmayer, P. Evolutionary Multiobjective Optimization for Base Station Transmitter Placement with Frequency Assignment. *IEEE Trans. Evolut. Comput.* **2003,** *7* (2), 189–203.
36. Kumar, R. V. R.; Gurugubelli, J. How Green the LTE Technology Can Be? In: *2011 2nd International Conference on Wireless Communication, Vehicular Technology,*

Information Theory and Aerospace and Electronic Systems Technology (Wireless VITAE), 2011; pp 1–5.
37. Richter, F.; Fehske, A. J.; Fettweis, G. P. Energy Efficiency Aspects of Base Station Deployment Strategies for Cellular Networks. In: *2009 IEEE 70th Vehicular Technology Conference Fall*, 2009; pp. 1–5.
38. Ajmone Marsan, M.; Chiaraviglio, L.; Ciullo, D.; Meo, M. Optimal Energy Savings in Cellular Access Networks. In: *2009 IEEE International Conference on Communications Workshops*, 2009; pp. 1–5.
39. Jardosh, Amit P., et al. Green WLANS: On-demand WLAN Infrastructures. *Mob. Netw. App.* **2009,** *14*.
40. Hauke Holtkamp, S. B.; Auer, G.; Haas, H. Minimizing Base Station Power Consumption. *IEEE J. Select. Areas Commun.* **2014,** *32*.
41. Heliot, F.; Imran, M. A.; Tafazolli, R. Energy Efficiency Analysis of Idealized Coordinated Multi-Point Communication System. In: *2011 IEEE 73rd Vehicular Technology Conference (VTC Spring)*, 2011; pp 1–5.
42. Han, T.; Ansari, N. On Greening Cellular Networks via Multicell Cooperation. *IEEE Wireless Commun.* **2013,** *20* (1), 82–89.
43. Willkomm, D.; Machiraju, S.; Bolot, J.; Wolisz, A. Primary User Behavior in Cellular Networks and Implications for Dynamic Spectrum Access. *IEEE Commun. Mag.*, *47* (3), 88–95.
44. Niu, Z.; Wu, Y.; Gong, J.; Yang, Z. Cell Zooming for Cost-Efficient Green Cellular Networks. *IEEE Commun. Mag.* **2010,** *48* (11), 74–79.
45. Yu, W. Multiuser Water-Filling in the Presence of Crosstalk. In: *2007 Information Theory and Applications Workshop*, 2007; pp 414–420.
46. Evangelinakis, D.; Sidiropoulos, N.; Swami, A. Joint Admission and Power Control Using Branch and Bound and Gradual Admissions. In: *2010 IEEE 11th International Workshop on Signal Processing Advances in Wireless Communications (SPAWC)*, 2010; pp 1–5.
47. Chen, T.; Kim, H.; Yang, Y. Energy Efficiency Metrics for Green Wireless Communications. In: *2010 International Conference on Wireless Communications and Signal Processing (WCSP)*, 2010; pp 1–6.
48. Jang, J.; Lee, K. B.; Lee, Y. H. Transmit Power and Bit Allocations for OFDM Systems in a Fading Channel. In: *GLOBECOM '03. IEEE Global Telecommunications Conference (IEEE Cat. No.03CH37489)* **2003,** *2*, 858–862.
49. Goudarzi, H.; Pakravan, M. R. Equal Power Allocation Scheme for Cooperative Diversity. In: *2008 4th IEEE/IFIP International Conference on Central Asia on Internet*, 2008; pp 1–5.
50. Lau, M. S. K.; Yue, W. Optimality and Feasibility of Equal Power Allocation of IDMA Systems. In: *2007 5th International Symposium on Modeling and Optimization in Mobile, Ad Hoc and Wireless Networks and Workshops*, 2007; pp 1–7.
51. Lin, X.; Andrews, J. G.; Ghosh, A.; Ratasuk, R. An Overview of 3GPP Device-to-Device Proximity Services. *IEEE Commun. Mag.* **2014,** *52* (4), 40–48.
52. Gandotra, P.; Jha, R. K. Device-to-Device Communication in Cellular Networks: A Survey. *J. Netw. Comput. App.* **2016,** *71*, 99–117. https://www.sciencedirect.com/science/article/pii/S1084804516301229

53. Wei, L.; Hu, R. Q.; Qian, Y.; Wu, G. Enable Device-to-Device Communications Underlaying Cellular Networks: Challenges and Research Aspects. *IEEE Commun. Mag.* **2014,** *52* (6), 90–96.
54. Gandotra, P.; Jha, R. K.; Jain, S. A Survey on Device-to-Device (D2D) Communication: Architecture and Security Issues. *J. Netw. Comput. App.* **2017,** *78*, 9–29. https://www.sciencedirect.com/science/article/pii/S1084804516302727
55. Improving Energy Efficiency, Low CO_2 Emission and TCO White Paper, Huawei Energy Efficiency Solution. Huawei Technologies Co. LTD. https://www.sciencedirect.com/science/article/pii/S1084804516302727
56. Feng, D.; Yu, G.; Xiong, C.; Y. Yuan-Wu, Li, G. Y.; Feng, G.; Li, S. Mode Switching for Energy-Efficient Device-to-Device Communications in Cellular Networks. *IEEE Trans. Wireless Commun.* **2015,** *14* (12), 6993–**7003**.
57. Jung, M.; Hwang, K.; Choi, S. Joint Mode Selection and Power Allocation Scheme for Power-Efficient Device-to-Device (D2D) Communication. In: *IEEE 75th Vehicular Technology Conference (VTC Spring)*, 2012; pp 1–5.
58. Mumtaz, S.; Saidul Huq, K. M.; Rodriguez, J.; Frascolla, V. Energy-Efficient Interference Management in LTE-D2D Communication. *IET Signal Process.* **2016,** *10* (3), 197–202.
59. Zhang, Y.; Yang, Y.; Dai, L. Energy Efficiency Maximization for Device-to-Device Communication Underlaying Cellular Networks on Multiple Bands. *IEEE Access* **2016,** *4*, 7682–**7691**.
60. Jiang, Y.; Liu, Q.; Zheng, F.; Gao, X.; You, X. Energy-Efficient Joint Resource Allocation and Power Control for D2D Communications. *IEEE Trans. Veh. Technol.* **2016,** *65* (8), 6119–**6127**.
61. Wei, L.; Hu, R. Q.; Qian, Y.; Wu, G. Energy Efficiency and Spectrum Efficiency of Multihop Device-to-Device Communications Underlaying Cellular Networks. *IEEE Trans. Veh. Technol.* **2016,** *65* (1), 367–380.
62. Lee, N.; Lin, X.; Andrews, J. G.; Heath, R. W. Power Control for D2D Underlaid Cellular Networks: Modeling, Algorithms, and Analysis. *IEEE J. Select. Areas Commun.* **2015,** *33* (1), 1–13.
63. Qiu, X.; Xuewen, L.; Dong, K.; Zhu, S. Energy efficiency analysis in device-to-device communication underlaying cellular networks. in *2013 IEEE 10th Consumer Communications and Networking Conference (CCNC)*, 625–630.
64. AlWreikat, L.; Chai, R.; Abu-Sharkh, O. M. Energy-Efficiency Based Resource Allocation for D2D Communication and Cellular Networks. In *2014 IEEE Fourth International Conference on Big Data and Cloud Computing*, 2014; pp 722–728.
65. Ali, M.; Qaisar, S.; Naeem, M.; Mumtaz, S. Energy Efficient Resource Allocation in D2D Assisted Heterogeneous Networks with Relays. *IEEE Access* **2016,** *4*, 4902–**4911**.
66. Yang, K.; Wu, J.; Gao, X.; Bu, X.; Guo, S. Energy-Efficient Power Control for Device-to-Device Communications with Max-Min Fairness. In *2016 IEEE 84th Vehicular Technology Conference (VTC-Fall)*, 2016; pp 1–5.
67. Xiao, X.; Tao, X.; Lu, J. A QOS-Aware Power Optimization Scheme in of DMA Systems with Integrated Device-to-Device (D2D) Communications. In: *2011 IEEE Vehicular Technology Conference (VTC Fall)*, 2011; pp 1–5.

68. Yu, C.-H.; Tirkkonen, O.; Doppler, K.; Ribeiro, C. Power Optimization of Device-to-Device Communication Underlaying Cellular Communication. In: *2009 IEEE International Conference on Communications*, 2009; pp 1–5.
69. Robat Mili, M.; Tehrani, P.; Bennis, M. Energy-Efficient Power Allocation in of DMA D2D Communication by Multi Objective Optimization. *IEEE Wireless Commun. Lett.* **2016,** *5* (6), 668–671.
70. Della Penda, D.; Fu, L.; Johansson, M. Energy Efficient D2D Communications in Dynamic TDD Systems. *IEEE Trans. Commun.* **2017,** *65* (3), 1260–**1273**.
71. Gupta, S.; Zhang, R.; Hanzo, L. Energy Harvesting Aided Device-to-Device Communication Underlaying the Cellular Downlink. *IEEE Access* **2017,** *5*, 7405–7413.
72. Ahmed, I.; Ikhlef, A.; Schober, R.; Mallik, R. K. Power Allocation for Conventional and Buffer-Aided Link Adaptive Relaying Systems with Energy Harvesting Nodes. *IEEE Trans. Wireless Commun.* **2014,** *13* (3), 1182–**1195**.
73. Sakr, A. H.; Hossain, E. Cognitive and Energy Harvesting-Based D2D Communication in Cellular Networks: Stochastic Geometry Modeling and Analysis. *IEEE Trans. Commun.* **2015,** *63* (5), 1867–**1880**.
74. 3rd Generation Partnership Project. Further Advancements for EUTRA Physical Layer Aspects. *IEEE Trans. Commun.* **2010,** Cedex, France, Tech. Rep. 36.814.
75. Yang, H. H.; Lee, J.; Quek, T. Q. S. Heterogeneous Cellular Network with Energy Harvesting-Based D2D Communication. *IEEE Trans. Wireless Commun.* **2016,** *15* (2), 1406–**1419**.
76. Maallawi, R.; Agoulmine, N.; Radier, B.; T. Ben Meriem. A Comprehensive Survey on Offload Techniques and Management in Wireless Access and Core Networks. *IEEE Commun. Surv. Tutor.* **2015,** *17* (3), 1582–**1604**.
77. Ge, X.; Tu, S.; Mao, G.; Wang, C.-X.; Han, T. 5G Ultra-Dense Cellular Networks. *IEEE Wireless Commun.* **2016,** *23* (1), 72–79.
78. Kamel, M.; Hamouda, W.; Youssef, A. Ultra-Dense Networks: A Survey. *IEEE Commun. Surv. Tutor.* **2016,** *18* (4), 2522–**2545**.
79. Ye, Y.; Zhang, H.; Xiong, X.; Liu, Y. Dynamic Min-Cut Clustering for Energy Savings in Ultra-Dense Networks. In: *2015 IEEE 82nd Vehicular Technology Conference (VTC2015-Fall)*, 2015; pp 1–5.
80. Bjornson, E.; Sanguinetti, L.; Kountouris, M. Deploying Dense Networks for Maximal Energy Efficiency: Small Cells Meet Massive MIMO. *IEEE J. Select. Areas Commun.* **2016,** *34* (4), 832–847.
81. Obaid, N.; Czylwik, A. Energy Efficiency Analysis of Dense Picocell Deployments. In: *OFDM 2014; 18th International OFDM Workshop 2014 (InOWo'14)*, 2014; pp. 1–6.
82. Yunas, S. F.; Valkama, M.; Niemela, J. Spectral and Energy Efficiency of Ultra-Dense Networks Under Different Deployment Strategies. *IEEE Commun. Mag.* **2015,** *53* (1), 90–100.
83. Yang, C. A Unified Design of Spectrum, Energy, and Cost Efficient Ultra-Dense Small Cell Networks. In *International Conference on Wireless Communications & Signal Processing, WCSP* **2015,** *Nanjing, China*, October 15–17, 2015. IEEE, **2015**; pp 1–4.
84. Su, L.; Yang, C.; L, C.-I. On Energy Efficiency and Spectral Efficiency Joint Optimization of Ultra Dense Networks. In *2015 IEEE Global Communications Conference (GLOBECOM)*, 2015; pp 1–6.

85. Yang, C.; Li, J.; Guizani, M. Cooperation for Spectral and Energy Efficiency in Ultra-Dense Small Cell Networks. *IEEE Wireless Commun.* **2016,** *23* (1), 64–71.
86. Samarakoon, S.; Bennis, M.; Saad, W.; Debbah, M.; Latva-aho, M. Ultra Dense Small Cell Networks: Turning Density into Energy Efficiency. *IEEE J. Select. Areas Commun.* **2016,** *34* (5), 1267–**1280**.
87. Su, L.; Yang, C.; L, C.-I. Energy and Spectral Efficient Frequency Reuse of Ultra Dense Networks. *IEEE Trans. Wireless Commun.* **2016,** *15* (8), 5384–**5398**.
88. Li, H.; Hu, D.; Chen, X.; Ci, S. High-Resolution Cell Breathing for Improving Energy Efficiency of Ultra-Dense Hetnets. In: *2015 IEEE Wireless Communications and Networking Conference (WCNC)*, 2015; pp 1458–**1463**.
89. Ghazanfari, A.; Tabassum, H.; Hossain, E. Ambient RF Energy Harvesting in Ultra-Dense Small Cell Networks: Performance and Trade-Offs. *IEEE Wireless Commun.* **2016,** *23* (2), 38–45.
90. Chen, L.; Yu, F. R.; Ji, H.; Rong, B.; Li, X.; Leung, V. C. M. Green Full-Duplex Self-Backhaul and Energy Harvesting Small Cell Networks with Massive MIMO. *IEEE J. Select. Areas Commun.* **2016,** *34* (12), 3709–**3724**.
91. Rusek, F.; Persson, D.; Lau, B. K.; Larsson, E. G.; Marzetta, T. L.; Edfors, O.; Tufvesson, F. Scaling up MIMO: Opportunities and Challenges with Very Large Arrays. *IEEE Signal Process. Mag.* **2013,** *30* (1), 40–60.
92. Zheng, K.; Zhao, L.; Mei, J.; Shao, B.; Xiang, W.; Hanzo, L. Survey of Large-Scale MIMO systems. *IEEE Commun. Surv. Tutor.* **2015,** *17* (3), 1738–**1760**.
93. Lu, L.; Li, G. Y.; Swindlehurst, A. L.; Ashikhmin, A.; Zhang, R. An Overview of Massive MIMO: Benefits and Challenges. *IEEE J. Select. Topics Signal Proces.* **2014,** *8* (5), 742–758.
94. Bjornson, E.; Sanguinetti, L.; Hoydis, J.; Debbah, M. Optimal Design of Energy-Efficient Multi-User MIMO Systems: Is Massive MIMO the Answer? *IEEE Trans. Wireless Commun.* **2015,** *14* (6), 3059–**3075**.
95. Larsson, E. G.; Edfors, O.; Tufvesson, F.; Marzetta, T. L. Massive MIMO for Next Generation Wireless Systems. *IEEE Commun. Mag.* **2014,** *52* (2), 186–195.
96. Van Chien, T.; Bjornson, E.; Larsson, E. G. Joint Power Allocation and User Association Optimization for Massive MIMO Systems, 2016.
97. Liu, W.; Han, S.; Yang, C. Energy Efficiency Scaling Law of Massive MIMO Systems. *IEEE Trans. Commun.* **2017,** *65* (1), 107–121.
98. Nguyen, T. M.; Ha, V. N.; L. Bao Le. Resource Allocation Optimization in Multi-User Multi-Cell Massive MIMO Networks Considering Pilot Contamination. *IEEE Access* **2015,** *3*, 1272–**1287**.
99. Lu, N.; Jiang, Y.; Zheng, F.; You, X. Energy Efficient Power Control for the Two-Tier Networks with Small Cells and Massive MIMO. In: *2016 IEEE Wireless Communications and Networking Conference*, 2016; pp 1–6.
100. Niyato, D.; Hossain, E. Dynamics of Network Selection in Heterogeneous Wireless Networks: An Evolutionary Game Approach. *IEEE Trans. Veh. Technol.* **2009,** *58* (4), 2008–**2017**.
101. Lee, B.; Choi, J.; Seol, J.-Y.; Love, D. J.; Shim, B. Antenna Grouping Based Feedback Compression for FDD-Based Massive MIMO Systems. *IEEE Trans. Commun.* **2015,** *63* (9), 3261–**3274**.

102. Gharagezlou, A. S.; Nangir, M.; Imani, N. Energy Efficient Power Allocation with Joint Antenna and User Selection in Massive MIMO Systems. *Computer Networks* **2022,** *216*, 1**09225**.
103. Al-Husseiny, Z.; Frenger, P. Enhancing LTE Energy Performance with Antenna Muting and Dynamic PSI-Omni Configuration. In: *2015 IEEE 81st Vehicular Technology Conference (VTC Spring)*, 2015; pp. 1–5.
104. Amirijoo, M.; Chai, Z.; Frenger, P.; Olin, B.; Moe, J. Self-Optimizing Antenna Muting—Energy Consumption and User Thoughput Analysis. In: *2012 International Symposium on Wireless Communication Systems (ISWCS)*, 2012; pp 46–50.
105. Goldsmith, A.; Jafar, S. A.; Maric, I.; Srinivasa, S. Breaking Spectrum Gridlock with Cognitive Radios: An Information Theoretic Perspective. *Proc. IEEE* **2009,** *97* (5), 894–914.
106. Irnich, T.; Kronander, J.; Li, G. Spectrum Sharing Scenarios and Resulting Technical Requirements for 5G Systems. In: *2013 IEEE 24th International Symposium on Personal, Indoor and Mobile Radio Communications (PIMRC Workshops)*, 2013; pp 127–132.
107. Kaushik, A.; Wunsch, F.; Sagainov, A.; Cuervo, N.; Demel, J.; Koslowski, S.; Jondral, F. Spectrum Sharing for 5G Wireless Systems (Spectrum Sharing Challenge). In: *2015 IEEE International Symposium on Dynamic Spectrum Access Networks (DySPAN)*, 2015; pp 1–2.
108. Baylis, C.; Fellows, M.; Barkate, J.; Tsatsoulas, A.; Rezayat, S.; Lamers, L.; Marks, R. J.; Cohen, L. Circuit Optimization Algorithms for Real-Time Spectrum Sharing Between Radar and communications. In: *2016 IEEE Radar Conference (RadarConf)*, pp. 1–4.
109. Zhai, C.; Liu, J.; Zheng, L. Cooperative Spectrum Sharing with Wireless Energy Harvesting in Cognitive Radio Networks. *IEEE Trans. Veh. Technol.* **2016,** *65* (7), 5303–**5316**.
110. Gao, H.; Ejaz, W.; Jo, M. Cooperative Wireless Energy Harvesting and Spectrum Sharing in 5G Networks. *IEEE Access* **2916,** *4*, 3647–3658.
111. Saha, S. K.; Ghoshal, S. P.; Kar, R.; Mandal, D. Cat Swarm Optimization Algorithm for Optimal Linear Phase Fir Filter Design. *ISA Trans.* **2913,** *52* (6), 781–794. https://www.sciencedirect.com/science/article/pii/S0019057813001055
112. Rao, J. B.; Fapojuwo, A. O. An Analytical Framework for Evaluating Spectrum/Energy Efficiency of Heterogeneous Cellular Networks. *IEEE Trans. Veh. Technol.* **2016,** *65* (5), 3568–3584.
113. Lu, W. D.; Gong, Y.; Ting, S. H.; Wu, X. L.; Zhang, N. T. Cooperative OFDM Relaying for Opportunistic Spectrum Sharing: Protocol Design And Resource Allocation. *IEEE Trans. Wireless Commu.* **2012,** *11* (6), 2126–2135.
114. Lu, W.; Wu, X.; Li, Q.; Zhang, N. Secondary Spectrum Access Based on Cooperative OFDM Relaying. In: *2012 IEEE 75th Vehicular Technology Conference (VTC Spring)*, 2014; pp 1–5.
115. Khoshkholgh, M. G.; Yamchi, N. M.; Navaie, K.; Yanikomeroglu, H.; Leung, V. C. M.; Shin, K. G. Radio Resource Allocation for OFDM Based Dynamic Spectrum Sharing: Duality Gap and Time Averaging. *IEEE J. Selected Areas Commun.* **2015,** *33* (5), 848–864.

116. O'Carroll, J.; Claussen, H.; Doyle, L. Partial GSM Spectrum Reuse for Femtocells. In: *2009 IEEE 20th International Symposium on Personal, Indoor and Mobile Radio Communications*, 2009; pp. 2111–2116.
117. Gong, X.; Ispas, A.; Dartmann, G.; Ascheid, G. Power Allocation and Performance Analysis in Spectrum Sharing Systems with Statistical CSI. *IEEE Trans.Wireless Commun.* **2013,** *12* (4), 1819–1831.
118. Liu, Y.; Liu, A.; Hu, Y.; Li, Z.; Choi, Y.-J.; Sekiya, H.; Li, J. Ffsc: An energy Efficiency Communications Approach for Delay Minimizing in Internet of Things. *IEEE Access* **2016,** *4*, 3775–3793.
119. Liu, C. H.; Fan, J.; Branch, J. W.; Leung, K. K. Toward QOI and Energy-Efficiency in Internet-of-Things Sensory Environments. *IEEE Trans. Emerg. Topics Comput.* **2014,** *2* (4), 473–487.
120. Zhang, D.; Zhou, Z.; Mumtaz, S.; Rodriguez, J.; Sato, T. One Integrated Energy Efficiency Proposal for 5G IoT Communications. *IEEE Internet of Things J.* **2016,** *3* (6), 1346–1354.
121. da Silva, F.; Lobao, G. et al. Particle Swarm Optimization Implementation for Minimal Transmission Power Providing a Fully-Connected Cluster for the Internet of Things. In: *2015 International Workshop on Telecommunications (IWT)*, 2015; pp 1–7.
122. Liguo, Q.; Jingkun, C.; Yourui, H.; Yiming, T. Power Control Algorithm of Internet-of-Things Node Based on Vague Set. In: *2010 Third International Symposium on Information Science and Engineering*, 2010; pp 251–254.
123. Bello, O.; Zeadally, S. Intelligent Device-to-Device Communication in the Internet of Things. *IEEE Syst. J.* **2016,** *10* (3), 1172–1182.
124. Zhou, Z.; Dong, M.; Ota, K.; Wang, G.; Yang, L. T. Energy-Efficient Resource Allocation for D2D Communications Underlaying Cloud-Ran-Based LTE—A Networks. *IEEE Internet of Things J.* **2016,** *3* (3), 428–438.
125. Zheng, T.; Qin, Y.; Zhang, H.; Kuo, S. Adaptive Power Control for Mutual Interference Avoidance in Industrial Internet-of-Things. *China Communications*, *13* (Supplement 1), 124–131.
126. Zhang, C.; Ahn, W.; Zhang, Y.; Childers, B. R. Live code update for iot devices in energy harvesting environments. In: *2016 5th Non-Volatile Memory Systems and Applications Symposium (NVMSA)*, **2016,** pp. 1–6.
127. Islam, K. R.; Tabassum, S.; Adhikary, T.; Razzaque, M. A. Parsimonious Renewable Energy Management Policies for Smart IoT Devices. In: *2016 5th International Conference on Informatics, Electronics and Vision (ICIEV)*, 2016; pp 758–763.
128. Lai, C.-F.; Lai, Y.-X.; Yang, L. T.; Chao, H.-C. Integration of IoT Energy Management System with Appliance and Activity Recognition. In: *2012 IEEE International Conference on Green Computing and Communications*, 2012; pp 66–71.
129. Yi, G.; Park, J. H.; Choi, S. Energy-Efficient Distributed Topology Control Algorithm for Low-Power IoT Communication Networks. *IEEE Access* **2016,** *4*, 9193–9203.
130. Saxena, N.; Roy, A.; Kim, H. Traffic-Aware Cloud Ran: A Key for Green 5G Networks. *IEEE J. Select. Areas Commun.* **2016,** *34* (4), 1010–1021.
131. Castelo Becerra, A.; Zeng, W.; Chow, M.-Y.; Rodra guez-Andina, J. J. Green City: A Low-Cost Testbed for Distributed Control Algorithms in Smart Grid. In: *IECON 2015—41st Annual Conference of the IEEE Industrial Electronics Society*, 2015; pp 001 948–001 953.

132. Wang, K.; Yu, J.; Yu, Y.; Qian, Y.; Zeng, D.; Guo, S.; Xiang, Y.; Wu, J. A Survey on Energy Internet: Architecture, Approach, and Emerging Technologies. *IEEE Syst. J.* **2018,** *12* (3), 2403–2416.
133. Galinina, O.; Pyattaev, A.; Johnsson, K.; Turlikov, A.; Andreev, S.; Koucheryavy, Y. Assessing System-Level Energy Efficiency Of MMwave-Based Wearable Networks. *IEEE J. Select. Areas Commun.* **2016,** *34* (4), 923–937.
134. Shokri-Ghadikolaei, H.; Boccardi, F.; Fischione, C.; Fodor, G.; Zorzi, M. Spectrum Sharing in MMwave Cellular Networks via Cell Association, Coordination, and Beam Forming. *IEEE J. Select. Areas Commun.* **2016,** *34* (11), 2902–2917.
135. International Energy Agency, Energy Efficiency Market Report, 2016.
136. Li, Y.; Jiang, T. Fixed-Point Algorithms for Energy-Efficient Power Allocation in Spectrum-Sharing Wireless Networks. In: *2016 IEEE Global Communications Conference (GLOBECOM)*, 2016; pp 1–6.
137. Majumder, T.; Mishra, R. K.; Sinha, A.; Singh, S. S.; Sahu, P. K. Congestion Control in Cognitive Radio Networks with Event Triggered Sliding Mode. *AEU—Int. J. Electron. Commun.* **2018,** *90*, 155–162.
138. Majumder, T.; Mishra, R. K.; Sinha, A.; Singh, S. S.; Sahu, P. K. Robust Congestion Control in Cognitive Radio Network Using Dynamic event-Triggered Sliding Mode. *Int. J. Commun. Syst.* **2019.**
139. Majumder, T.; Mishra, R. K.; Singh, S. S.; Sahu, P. K. Robust Congestion Control in Cognitive Radio Network Using Event Triggered Sliding Mode Based on Reaching Laws. *J. Franklin Insti.* **2020,** *357* (11), 7399–7422.
140. Majumder, T.; Singh, S. S.; Sahu, P. K.; Sinha, A.; Mishra, R. K. Robust Nonlinear Congestion Controller for Time Delayed and Uncertain Cognitive Radio Based Wireless Network. In: Proceedings of the 2015 *IEEE Power, Communication and Information Technology Conference (PCITC)*, **2015,** pp 248–254.
141. Majumder, T.; Mishra, R. K.; Sinha, A.; Singh, S. S.; Sahu, P. K. Robust Nonlinear Congestion Controller for Cognitive Radio Based Wireless Network. In: *IEEE Sponsored 2nd International Conference on Innovations in Information, Embedded and Communication Systems, ICIIECS'15*, 2015.
142. Majumder, T.; R.Mishra, K.; Sinha, A.; S.Singh, S.; Sahu, P. K. Congestion Control in Cognitive Radio Networks Using Fractional Order Rate Reaching Law Based Sliding Modes. In: *Proceedings of the Foundations and Frontiers in Computer, Communication and Electrical Engineering, Taylor and Francis*, **2016,** pp 219–224.

CHAPTER 3

Secure Adaptive Cluster-Based Routing Using Modified Cuckoo Search Algorithm for Health Monitoring Systems in Smart Cities

M. S. SOWMYASHREE[1], I. G. SARITHA[1], I. G. NAVEEN[2], and C. S. MALA[3]

[1]*Department of ETE, BMS Institute of Technology & Management, Bengaluru, India*

[2]*Department of ECE, Nitte Meenakshi Institute of Technology, Bengaluru, India*

[3]*Department of ECE, BMS Institute of Technology & Management, Bengaluru, India*

ABSTRACT

WSNs (wireless sensor networks) are used in numerous functions in agriculture, military surveillance, medical and clinical care, ecological inspection, mechanical gathering, disaster recovery tasks, farming, and many others. All these applications are essential in smart cities, which provide security in monitoring. Monitoring in the field of medical assistance can provide remote protection in the absence of human interference. To obtain this, the sensor nodes' aid is of great importance. To link with the other nodes in the network, the WSN is organized in a random order. As WSN nodes are battery-powered, energy becomes one of the important primary constraints of WSNs. Furthermore, malicious attacks (e.g., a black hole attack) may damage the nodes, resulting in various network security

5G Green Communication Networks for Smart Cities. Devasis Pradhan, Rajeswari, Hla Myo Tun, Naw Khu Say Wah, & Thandar Oo (Eds.)

difficulties. This attack may modify the patient-sensed data and send the wrong data to the hospital monitoring system, which causes an incorrect diagnosis of a patient and leads to severe issues. A concept known as Secure Adaptive Cluster-Based Routing is necessary for utilizing the Modified Cuckoo Search Algorithm (SACRMCSA) to overcome the security and energy challenges in WSN. The clustering process in the proposed work is carried out employing the K-means clustering process.

The modified Cuckoo Search Optimization (CSO) method is used to identify an optimal Cluster Head (CH). Then CSO is used in conjunction with the AODV protocol, which achieves an ideal pathway for secure routing. The primary goal of this work is to offer an algorithm that increases WSN energy efficiency while simultaneously providing secure communication so that the data is not altered and correct data reaches the hospital management system and the patient is diagnosed correctly without any faults. Associated with the prevailing Secured Routing-Based Protocol on the Multiobjective Ant Colony Optimization (SRPMA) approach, the SACRMCSA method delivers packets at a rate of 95.91%, reducing the packet loss and providing security by overcoming the malicious nodes and increasing the energy life of the node. Such monitoring systems are essential in intelligent cities so that proper management is achieved in terms of remote healthcare, which reduces unnecessary moving of patients on the road to a hospital, reducing traffic and pollution.

3.1 INTRODUCTION

A municipality is considered a "smart city" if it makes use of information and communication technology (ICT) to improve administrative efficiency, enlighten the public, raise the bar for public services, and improve the welfare of its citizens.[24] The adoption of smart cities is being fueled by new trends like automation, machine learning, and healthcare using WSN. Given the growing requirement for intelligent, remote, and real-time healthcare services in smart cities, automated intelligent healthcare observation is required to deliver more comprehensive treatment to inhabitants. To meet the demand for high-quality treatment, health-related media or signals collected from smart devices or objects are sent and analyzed as part of this monitoring. It is difficult to develop an approach for handling healthcare data analytics or indications in the remote monitoring system;[25] therefore, an architecture used to create an intelligent healthcare system is shown in Figure 3.1 below.

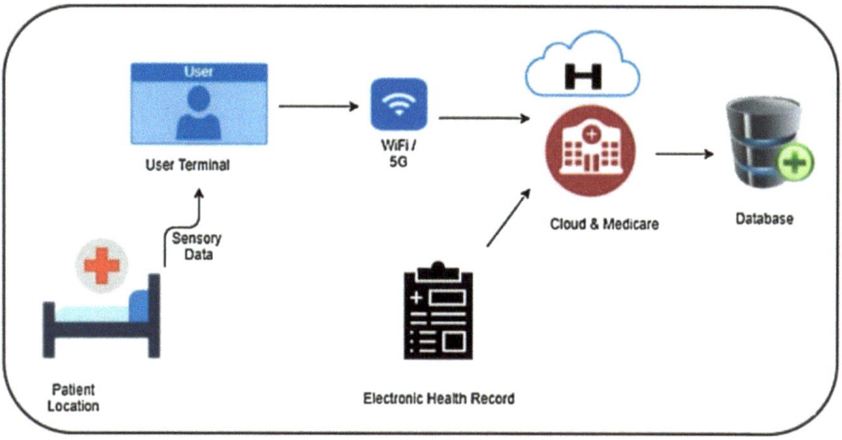

FIGURE 3.1 Architecture of a smart healthcare monitoring system.

WSN combines tiny, precise sensor focus points that enable many activities, including discovery, broadcast, and managing various distant communication chores.[1,2] A structured network in a WSN is a collection of sensors in ad hoc form. In an unstructured WSN network, the sensor centers are positioned in a predetermined manner.[3] The sensor hubs use the two-way radio frequency to communicate across an air channel.[4] They collect and store important records for further crucial tasks that the sink hub handles.[5] Moving forward, the sensor hubs must be used continuously or on an as-needed basis for quick data collection and authentication in various circumstances.[6] Power necessitates attention for efficient data exchange in the system because of the limited battery capacity.[7] As a result, the clustering and forwarding approach lowers the framed network system's processing overhead and energy consumption. However, the routing strategy has many security issues with the route taken to get to the destination from the source point.[8,9]

Recent events have shown that WSN is susceptible to multiple assaults, which makes data transmission less secure. The DoS attack, black hole assault, Sybil attack, and spoofing attack are just a few of the attacks that hackers have carried out on sensor nodes.[10,11] As a result, hackers can affect the sensor nodes and cause them to behave maliciously. The network's cancerous nodes make the data transmission process insecure for the entire network.[12] Unauthorized data access becomes conceivable, which makes it incredibly challenging to recover any crucial data gathered at the nodes.[13–15] This results in developing a safe and energy-efficient routing protocol for the network system and supporting security. The following elements must be considered while developing a protocol to achieve secured message communication over WSN: dependability, forward confidentiality, backward confidentiality, originality, nonrepudiation, and accessibility.[16] As a result, a secure adaptive cluster-based routing protocol with a modified cuckoo search algorithm (SACRMCSA) is suggested in this paper. SACRMCSA combines an AODV routing protocol with a cuckoo search optimization method. The following list summarizes the investigation's principal findings:

- Initially, the K-means algorithm for clustering is applied for an efficient clustering operation of the nodes available in the network.
- The modified cuckoo search optimization algorithm is introduced for an optimal selection of the CH. The choice of the CH is accomplished based on the calculation of fitness functions, which include node degree, residual energy, trust, and value distance.
- In addition, the AODV routing protocol is integrated with the cuckoo search algorithm to create the best possible path for safe data transfer in the network.

The offered work's entire composition is laid down in paper format as follows: The associated results are evaluated in Section 2. In Part 3, the SACRMCSA is explained in detail. Section 4 of the SACRMCSA provides an explanation of the performance analysis. The paper is concluded in Section 5.

3.2 LITERATURE REVIEW

The MAC Centralized Routing Protocol (MCRP), created by Ahutu and El-Ocla,[17] was used to implement MCRP and create a centralized network intelligence. To efficiently detect wormhole attacks, the MCRP technique uses a Base Station (BS) component that uses less power while maintaining harmony between the BS and sensor hubs. Even though the MCRP protocol increased the number of nodes to its maximum, the WSN's simulation performance improved significantly. However, MCRP needed to be upgraded to improve WSN performance when fewer nodes were present. Throughput, frame rate, and other metrics were used to assess the MCRP method's output.

An RBMSC mode is a reputation-based mechanism to encourage cooperation created by Shi et al.[18] This approach uses a developed protocol to adopt a path for data transport. This newly created protocol offers protection against attacks launched by malicious nodes. The suggested system can create a secure way of transmission using the Dijkstra algorithm. This RBMSC mode has a limitation that makes it impossible to quantify the time delay during data transfer. The results are assessed regarding the packet loss rate, number of nodes, and packet delivery ratio.

To improve the dynamic source routing (DSR) protocol's quality of service, Salari-Moghaddam et al.[19] designed a routing algorithm based on trust values. This technique employs a trust-based DSR (TDSR) to block rogue nodes from entering the transmission route. Additionally, the Enhanced DSR (EDSR) approach was applied, raising the service standard. Due to this inclusion, the DSR approach enabled improved execution in terms of packet delivery ratio, which decreased the network's energy consumption. The proposed DSR and TDSR protocols performed less well to find the shortest way; occasionally, the proposed approach also chose the longer path since it consumed more energy.

Sun et al. put forth a secured routing-based protocol on multiple objective ant colony optimization (SRPMA) technique for WSN.[20] This method considered two objective functions. By taking into account the typical residual energy of the routing channel, the aim's initial goal is to use as little energy as possible. The following goal function was to consider the specific value of the trust in the routing path, ensuring that all route nodes were outfitted with security. In contrast to a black hole assault on a WSN, this SRPMA implementation method produced a good performance. However, the SRPMA approach had a flaw that made it take longer to anticipate the network's level of energy usage.

The adaptive sunflower optimization (ASFO) algorithm is presented in this study to increase the coverage area and decrease the issue of hole attacks in the network. An artificial immune-based routing protocol is used to discover the most effective routing path that can convey data from the CH to the BS. The quality of service could not be improved due to network overload, which is a restriction. The suggested model's outcomes were estimated in terms of energy consumption, delay, packet delivery ratio, and throughput.[21]

3.3 SACRMCSA SYSTEM

Most of the existing wireless sensor networks consume more energy as they are used in a variety of applications, and there is more delay in transmitting the data from source to destination depending on the number of clients. This reduces the overall efficiency of the network. For better utilization of resources, adaptive network management and security are important primary factors. To obtain this, the Cluster Head (CH) selection phase, followed by the Routing Operations phase, makes up the initial phases of the SACRMCSA method. Initially, K-means, a clustering technique, is used to complete the clustering process. Following clustering, the modified cuckoo search optimization (CSO) algorithm is used to conduct the selection of CH and routing. In the planned SACRMCSA technique, the CH is selected based on the restrictions known as the fitness function. The four function parameters—node degree, distance, trust value, and residual energy—are used to evaluate the fitness function. The node trust value is calculated to prevent black hole attacks from entering the system network. The method is known as adaptive clustering as the harmful nodes are avoided during the selection of CH. The main goal of the SACRMCSA approach is to provide secure clustering to increase the network's life. Figure 3.2 displays the SACRMCSA flow diagram.

3.3.1 K-MEANS CLUSTERING ALGORITHM

This is a split-based categorization system, as seen in Figure 3.3. The n samples, represented by the number n, are divided into k subgroups in this technique. High comparisons within the cluster are caused by the subsets that take the shape of groups. In the beginning, samples of k are carefully chosen at random from the data set and then models are calculated using the Euclidean distance formula between the various K centers. The new clustering center is determined after computing the subset mean. The actions are combined with the K-means clustering technique to establish a new cluster center.

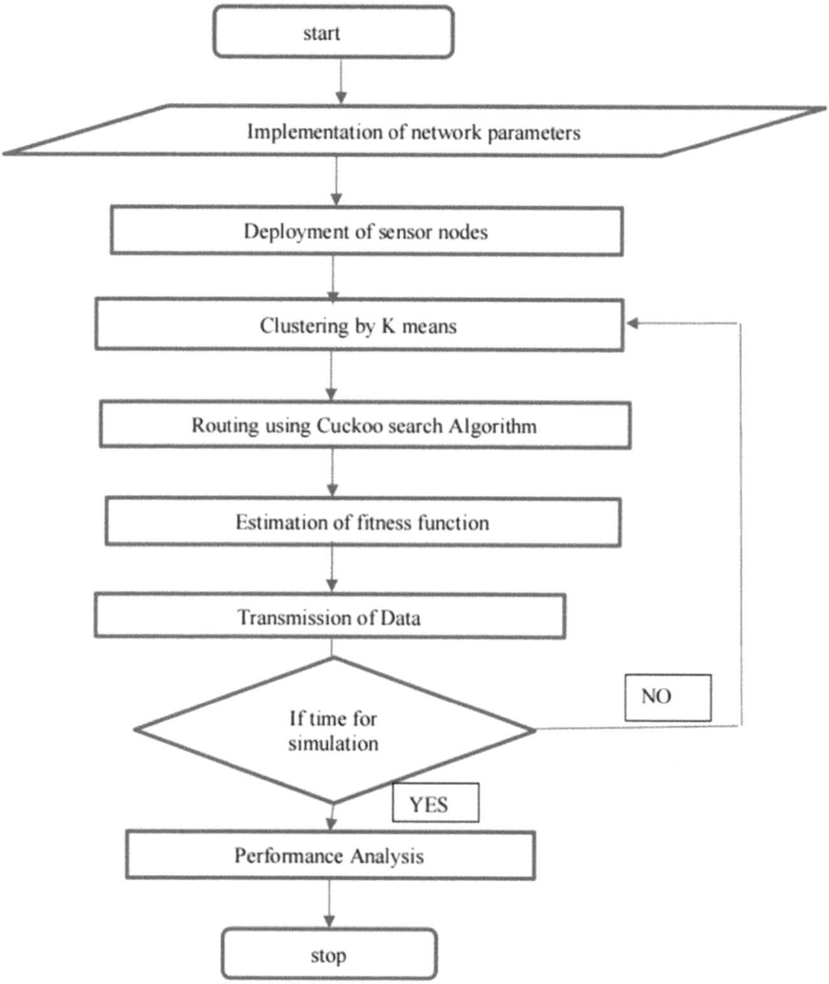

FIGURE 3.2 Flow chart of the SACRMCSA method.

3.3.2 CUCKOO SEARCH ALGORITHM OVERVIEW

To solve optimization-related issues, the CS algorithm, a powerful and efficient heuristic tool, simulates the parental parasite and the Levy flight of the cuckoo. The cuckoo bird is distinguished because it does not construct its own nest during the breeding season. To reproduce, it swaps

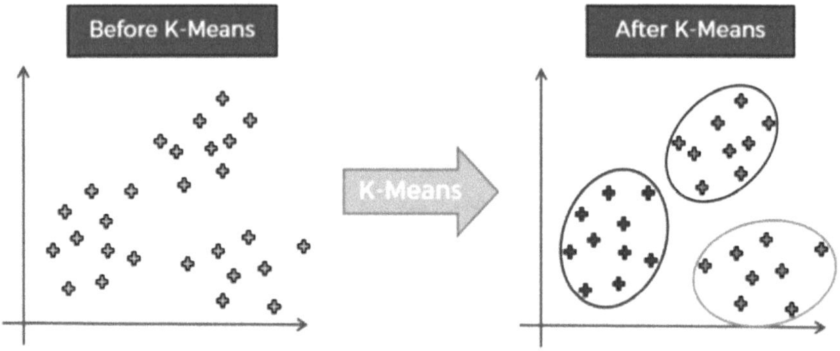

FIGURE 3.3 K-means clustering algorithm.

Source: Reprinted from Medium. https://medium.datadriveninvestor.com/k-means-clustering-4a700d4a4720

out its original eggs for those laid by other birds. After the eggs hatch, the cuckoo links the characteristics of its breed together. The route for constructing the new nest site for breeding is then assessed using eq. (3.1), as shown below:

$$x_i^{(t+1)} = x_i^{(t)} + \propto \oplus Levy(\lambda); i = 1,2,\ldots n-1 \tag{3.1}$$

The term $x_i^{(t)}$ denotes the location of the nest of ith bird in t generation and α signifies the step size control. $Levy(\lambda)$ stands for Levi's random path search, and it is attained by the following eq. (3.2).

$$Levy(\lambda) = t^{-\lambda}; 1 < \lambda < 3 \tag{3.2}$$

3.3.2.1 MODIFIED CUCKOO SEARCH-BASED CLUSTER HEAD SELECTION

In this study, the modified cuckoo search optimization algorithm selects the best sensor node to serve as the CH for each cluster. Determining the optimal CH increases the network's longevity. Data aggregation and cluster management are under the purview of the CH; therefore, choosing the right node from among all cluster members becomes crucial. Then, the CH is carefully selected depending on the fitness functions, including trust value, residual energy distance, and node degree.

3.3.2.1.1 Estimation of Fitness Function for Selection of Cluster Head

By considering the optimal CH in each cluster, the following fitness metrics were considered for choosing CH: node degree, trust value, residual energy, and distance. Below is a list of all the terms and their definitions.

a. Residual Energy

Each cluster has a cluster head, which carries out a variety of tasks such as data aggregation and cluster management. Given its significant responsibilities, the CH should possess high levels of residual energy and the ability to balance energy use among all the clusters. While calculating the energy known as residual energy, the following equation must be used (eq. 3.3).

$$F_1 = \sum_{i=1}^{m} \frac{1}{RE_{CH_i}} \tag{3.3}$$

Here, the residual energy in a node is represented as RE_{CH}, and the node number is denoted as m.

b. Node Degree

The hop count associated with each node specifies node degree, a vital element that must be counted during the next-hop choice. If the subsequent hop is chosen with the lowest node degree, the related node will last longer. The equation (3.4) is used to calculate the node degree.

$$F_2 = \sum_{i=1}^{m} |CM_i| \tag{3.4}$$

Here, CM_i is a representation of the number of cluster nodes in the ith cluster.

c. Distance

The distance between each cluster member and CH is determined by eq. (3.5). The computation of length in a group is accomplished to enhance the link quality between the nodes.

$$F_3 = \sum_{j=1}^{m} \left[\frac{\sum_{i=1}^{|CM_j|} d(CH_j, CM_i)}{|CM_j|} \right] \quad (3.5)$$

Here, the Euclidean distance is used among CH of jth cluster, and Cluster Member of the jth cluster is symbolized a s $d(CH_j, CM_i)$.

d. Trust Value

A crucial consideration in the design of a secure system is the trust value calculation. The computation of the trust value is used for trust routing, access control, and authentication of that node. The calculation of the trust value is based on eq. (3.6).

$$F_4 = \frac{DS_{i,j}(t)}{DR_{i,j}(t)} \quad (3.6)$$

The term $DS_{i,j}(t)$ indicates the time taken by the transmitter node to deliver the data, and $DS_{i,j}(t)$ $DR_{i,j}(t)$ is the time needed by the official receiver to accumulate data.

3.3.2.1.2 *Initialization*

The values of each node's fitness function are calculated during the initialization phase using the corresponding equations. The trust values must be carefully evaluated to keep malicious nodes from becoming the CH. The sensors are arranged in decreasing order based on fitness function. The top 20% of sorted nodes are carefully chosen as truly qualified CHs after receiving a list of nodes in that category. The nodes in these clusters are the eggs, and the clusters are host nests. After the initialization step is finished, the modified cuckoo search iterative aspect method is started.

3.3.2.1.3 *Iterative Process*

All host nests have access to the most suitable CH that was chosen. There is a potential that more than one of the top CH candidates will be found in one of the clusters. In these cases, the best CH for that cluster is selected by matching the fitness function values of the CH members with those of other

CH members in the same host nest. Up until the best CH is established for every host nest in the network, this iterative element will keep happening.

3.3.2.2 MODIFIED CUCKOO SEARCH-BASED ROUTING ALGORITHM

Modified Cuckoo Search is used to pick the best route for data transfer. The trust value in the nodes is calculated to prevent malicious nodes such as black hole attacks. The likelihood of a link failure can be decreased by avoiding hostile nodes while choosing the path.

3.3.2.2.1 Initialization

In this phase, the cluster bunch is known as the host nest, and sensor nodes are considered eggs. The location of each egg is initiated by an arbitrary id known as node_id, which has a random value ranging from 1 to n. Let us consider $E_i = (E_i^1, E_i^2 \ldots\ldots\ldots E_i^m)$ the ith egg, and each egg position as $E_i^d, where 1 \leq d \leq m$ which characterizes the node_id ranging from 1 to n. The probable best route is found between the nodes. The fitness functions of each node in the path are computed. Once the initialization phase is completed, the iteration is then initiated.

3.3.2.2.2 Process of Iteration

The calculated probable fitness functions of nodes are then used for the iterative aspect to discover the best possible path for the transmission of the data. The distance amongst the nodes is computed with the help of fitness functions, reducing link breakdown during communication. This leads to an increase in the packet delivery ratio. In the constructed values of fitness function in the modified cuckoo search algorithm, the optimum nodes are carefully chosen. 3.2.1.1 discusses the fitness function calculation for forwarding and sending.

3.3.2.2.3 Iterative Process

The iterative process uses probable estimated values of the fitness functions to find the most effective route for data transfer. The distance amongst the nodes is computed with the help of fitness functions, reducing link breakdown during communication. This leads to increases in the packet delivery ratio of the network. With the constructed values of the fitness function in the modified cuckoo search algorithm, the optimum nodes are carefully chosen.

3.3.2.2.4 Routing Based on AODV

To establish a path after obtaining the fitness function values, the routing protocol known as AODV is combined with a modified cuckoo search algorithm in the proposed study. In this method, the node, the source, broadcasts the message called a route request and sends node_id and fitness values. For request_message sent, a reply message known as a route reply is obtained. The source node chooses the best path among the received responses by evaluating the fitness values attached with the route reply message.

3.4 DISCUSSION ON SIMULATION RESULTS

The Network Simulator (NS) version 2.34 platform simulates the random nodes in an area of 1200 m × 1200 m for the SACRMCSA method. The SACRMCSA simulation is examined, considering genuine black hole attack data. Black hole attacks turn some of the nodes into cancerous nodes. The node energy of the sink is vast, and the initial energy of the sensor is stable. The parameters of the nodes are shown in Table 3.1 below.

3.4.1 PERFORMANCE EVALUATION

The parameters used to evaluate the performance of the SACRMCSA method are examined as follows:

3.4.1.1 END-TO-END DELAY (EED)

The first parameter, EED, is defined as the total amount of time needed by the receiver to process the ratio of packets transmitted to packets received. As can be seen below, an equation is used to determine the EED (eq. 3.7).

$$EED = \frac{time\,taken\,by\,the\,receiver\;for\;packet\;transmission}{The\,number\;of\;packets\,received\;by\,the\;receiver} \tag{3.7}$$

TABLE 3.1 Parameter Specifications.

Parameter considered	Value
MAC protocol	802.11/MAC
Pattern of Antenna	Omni
Radius of Communication	300 m
Initial energy	75 J
Data speed	450 kbit/s
Queue type	PriQueue
Interface type of the network	WirlessPhy
Wireless propagation	TwoRayGround
Simulation time	100 s
Size of packets	210 Bytes
Number of nodes	**125**
Number of connections	20

As calculated in Figure 3.4, the EED of the suggested approach, SACRMCSA, is contrasted with the current procedure, SRPMA[20]. The graph demonstrates that the suggested SACRMCSA method's EED is considerably lower than the SRPMA[20] method now in use. The fitness function values are evaluated for data transmission to obtain the nodes' distances. From the obtained values, EED decreased with the proposed method.

3.4.1.2 PACKET DELIVERY RATE (PDR)

The second indicator, PDR, shows the proportion of successfully delivered packets to all packets sent during data transmission. The PDR is determined as follows using eq. (3.8):

$$PDR = \frac{T_{ND}}{T_{NS}} \tag{3.8}$$

Here, T_{ND} is the number of packets delivered, and the number of packets sent is denoted by T_{NS}.

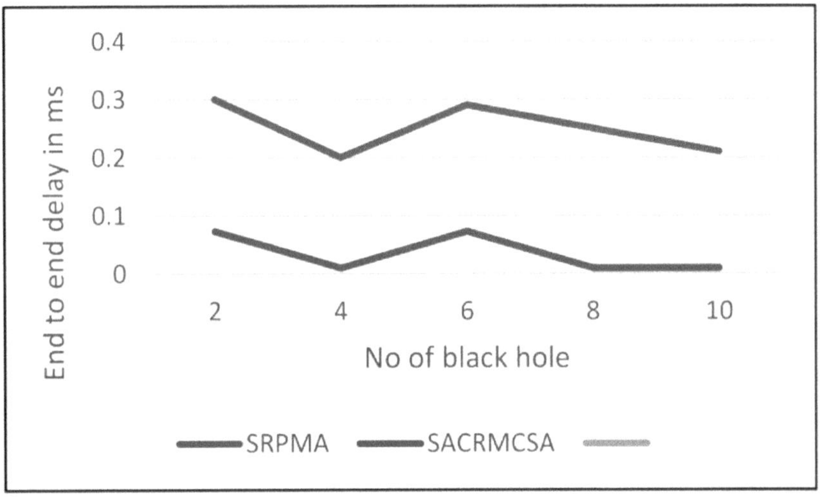

FIGURE 3.4 Graphical comparison of end-to-end delay.

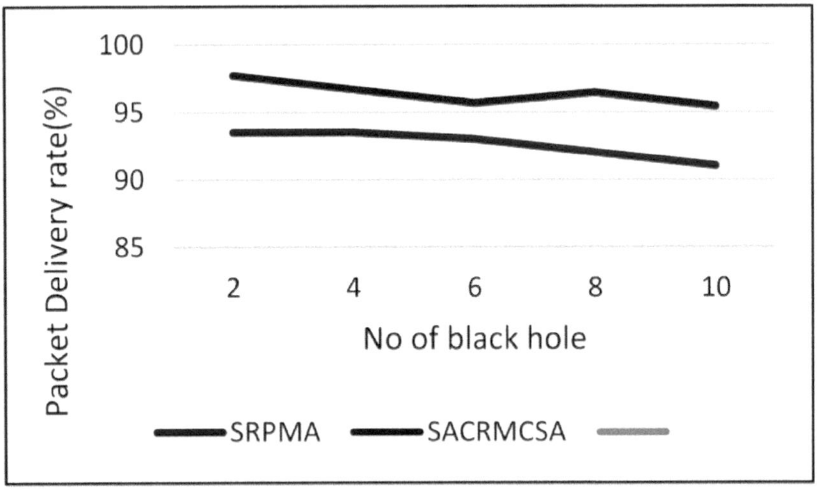

FIGURE 3.5 Graphical comparison of PDR.

The comparison of the PDR of the proposed structure SACRMCSA with SRPMA[20] is illustrated in Figure 3.5. In comparison to the currently used approach, SRPMA, the graph demonstrates that the SACRMCSA method has a maximum packet delivery rate.[20] The nodes' trust values are contained in the fitness function, which lowers the likelihood that links will break during communication. This raises the SACRMCSA method's pack delivery rate.

3.4.1.3 PACKET LOSS RATE (PLR)

The third parameter, PLR, is obtained by calculating the total number of data packets delivered by the source nodes and the total number of data packets received by the sink throughout the simulation phase. In order to calculate the packet loss rate, eq. (3.9) is employed.

$$Packet\,loss\,rate = 1 - \frac{No.of\ received\ data\ packets\ from\ the\ sink}{No.of\ sent\ data\ packets\ by\ the\ source\ node} \tag{3.9}$$

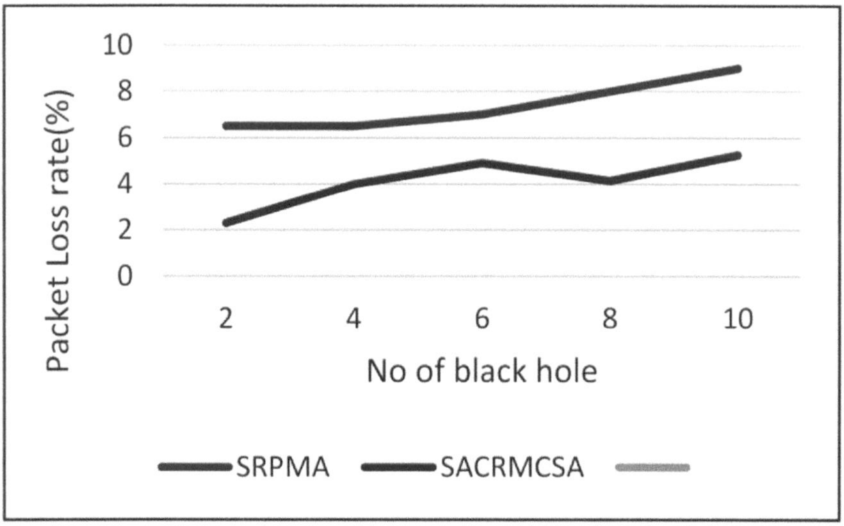

FIGURE 3.6 Graphical comparison of packet loss rate.

Figure 3.6 shows the difference between SACRMCSA and the existing SRPMA[20] method in terms of PLR. The proposed SACRMCSA PLR is less matched to the existing SRPMA[20] method. To reduce the link failure, trust values and distance are used to calculate the fitness function, which reduces the failure of the link. The routing protocol AODV is used to avoid a link break during the data transmission so that an alternate path is chosen. Henceforth, the rate of packet loss in the SACRMCSA method is minimized.

3.4.1.4 ROUTING LOAD (RL)

The ratio of the number of data packets delivered to those that were transmitted is the fourth parameter, RL. Equation (3.10) yields the results for the routing load as follows:

$$Routing\, load = \frac{\delta}{\chi} \tag{3.10}$$

Here, δ is the quantity of data packets transferred, and the delivered quantity of data packets is symbolized as χ.

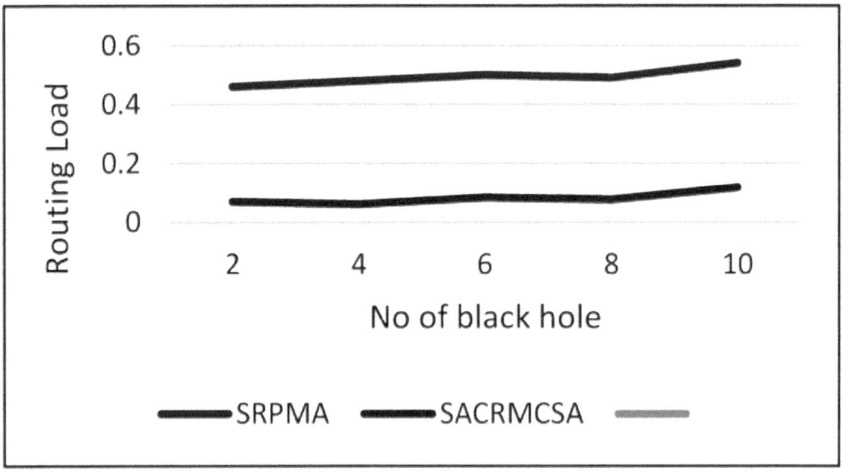

FIGURE 3.7 Graphical comparison of routing load.

Figure 3.7 compares the available SRPMA[20] methodology and the proposed SACRMCSA method's routing load evaluation. By minimizing the control messages sent throughout the path selection phase, the RL of the suggested SACRMCSA approach is decreased. The use of routing protocols such as AODV reduces the control messages effectively. The proposed SACRMCSA method achieves a lesser routing load when evaluated with the SRPMA[20] approach.

3.4.1.5 AVERAGE ENERGY CONSUMPTION

The fifth parameter, referred to as average energy consumption, is the total energy received by the node and energy transmitted. The calculation of energy consumption is evaluated using eq. (3.11) as shown below:

$$Average\,energy\,consumption = (E_{RX} \times number\ of\ nodes) + E_{TX} \quad (3.11)$$

Here, E_{RX} is the energy received, and the energy that is transmitted to the nodes is characterized as E_{TX}.

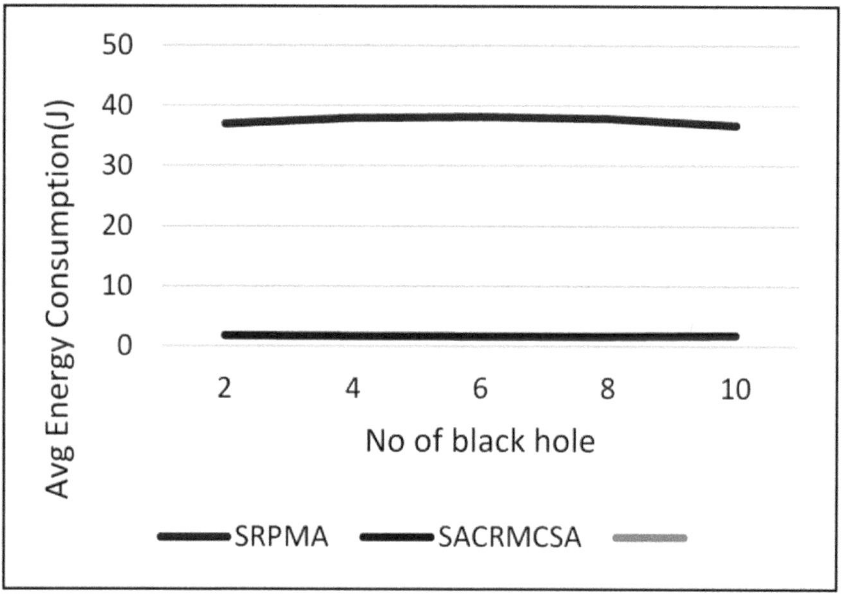

FIGURE 3.8 Graphical representation of average energy consumption.

Figure 3.8 compares the SACRMCSA method's average energy consumption to the available SRPMA[20] technique. The proposed SACRMCSA method signifies a more significant energy consumption difference. The Euclidean distance is computed for the path discovery process to find the closest node to transmit the data. Due to this process, the consumption of energy has decreased. Table 3.2 below affords a comparative study of the proposed SACRMCSA technique and SRPMA[20] method in terms of the evaluation parameters, end-to-end delay, packet loss rate, and average energy consumption. Similarly, Table 3.3 compares the proposed SACRMCSA method with the SRPMA[20] regarding packet delivery rate and routing load.

TABLE 3.2 Comparison of the SACRMCSA Method's End-to-End Delay, Average Energy Usage, and Packet Loss Rate with SRPMA[20].

Number of black hole nodes	Average energy consumption (J)		EED (ms)		PLR (%)	
	SRPMA[20]	SACRMCSA	SRPMA[20]	SACRMCSA	SRPMA[20]	SACRMCSA
2	37	1.76	0.3	0.0732	6.5	2.2659
4	38	1.67	0.2	0.0096	6.5	3.9742
6	38.2	1.63	0.29	0.0728	7	4.9193
8	37.9	1.64	0.25	0.0096	8	4.1433
10	36.8	1.79	0.21	0.0095	9	5.2609

TABLE 3.3 Analysis of the SACRMCSA Method's Routing Burden and Packet Delivery Rate in Comparison to the SRPMA[20] Technique.

Number of black hole nodes	RL		PDR (%)	
	SRPMA[20]	SACRMCSA	SRPMA[20]	SACRMCSA
2	0.46	0.067	93.5	97.7051
4	0.48	0.061	93.5	96.6839
6	0.5	0.084	93	95.651
8	0.49	0.075	92	96.4386
10	0.54	0.117	91	95.4185

3.5 CONCLUSIONS

In the proposed research work, the clustering algorithm known as K-means is employed to make the clustering task of the nodes available in a random manner in a Wireless Sensor Network. The Cuckoo Search Optimization process chooses an optimum CH in each cluster present.

Furthermore, the CSO is combined with the routing protocol known as AODV to obtain the best pathway for data transmission. The CH selection and routing functions are accomplished and centered on four different fitness functions in this research work. The fitness function parameters are the node degree, residual energy, trust value, and distance calculation. To prevent the selection of malicious nodes from becoming CHs during the transmission phase, the trust values are involved in the fitness value calculation, which prevents the nodes from getting impacted by the attack known as a black hole.

As a result, the network delivers secured data transmission using the proposed SACRMCSA mechanism. When compared to the current SRPMA approach, the SACRMCSA method performs significantly better when comparing simulated and implemented results. The energy consumption value of SACRMCSA is 1.14 J for eight black hole nodes, which is less than the current SRPMA method. Future WSN performance improvements will be achieved by the fusion of optimization techniques.

KEYWORDS

- **cuckoo search algorithm**
- **WSN**
- **AODV**
- **SACRMCSA**

REFERENCES

1. Guleria, K.; Verma, A. K.; Meta-Heuristic ant Colony Optimization Based Unequal Clustering for Wireless Sensor Network. *Wireless Pers. Commun.* **2019,** *105* (3), 891–911.
2. Mittal, N.; Moth Flame Optimization-Based Energy Efficient Stable Clustered Routing Approach for Wireless Sensor Networks. *Wireless Pers. Commun.* **2019,** *104* (2), 677–694.
3. Selvi, M.; Thangaramya, K.; Ganapathy, S.; Kulothungan, K.; Nehemiah, H. K.; Kannan, A.; An Energy-Aware Trust Based Secure Routing Algorithm for Effective Communication in Wireless Sensor Networks. *Wireless Pers. Commun.* **2019,** *105* (4), 1475–1490.

4. Kalidoss, T.; Rajasekaran, L.; Kanagasabai, K.; Sannasi, G.; Kannan, A.; QoS Aware Trust-Based Routing Algorithm for Wireless Sensor Networks. *Wireless Personal Commun.* **2010,** *110* (4), 1637–1658.
5. Sharma, R.; Vashisht, V.; Singh, U. Fuzzy Modelling-Based Energy Aware Clustering in Wireless Sensor Networks Using Modified Invasive Weed Optimization. *J. King Saud Univ.—Comput. Inf. Sci.* **2019**.
6. Sharma, R.; Vashisht, V.; Singh, U. EE TMFO/GA: A Secure and Energy Efficient Cluster Head Selection in Wireless Sensor Networks. *Telecommun. Syst.* **2020,** 1–16.
7. Sahoo, B. M.; Amgoth, T.; Pandey, H. M. Particle Swarm Optimization Based Energy Efficient Clustering and Sink Mobility in Heterogeneous Wireless Sensor Network. *Ad Hoc Networks* **2020,** *106*, 102237.
8. Famila, S.; Jawahar, A. Improved Artificial Bee Colony Optimization-Based Clustering Technique for WSNs. *Wireless Personal Commun.* **2020,** *110* (4), 2195–2212.
9. Christopher, V. B.; Jasper, J. Jellyfish Dynamic Routing Protocol with Mobile Sink for Location Privacy and Congestion Avoidance in Wireless Sensor Networks. *J. Syst. Architect.* **2021,** *112*, 10**1840**.
10. Khan, T.; Singh, K.; Abdel-Basset, M.; Long, H. V.; Singh, S. P.; Manjul, M. A Novel and Comprehensive Trust Estimation Clustering Based Approach for Large Scale Wireless Sensor Networks. *IEEE Access* **2019,** *7*, 58221–**58240**.
11. Dhand, G.; Tyagi, S. S. SMEER: Secure Multi-Tier Energy Efficient Routing Protocol for Hierarchical Wireless Sensor Networks. *Wireless Pers. Commun.* **2019,** *105* (1), 17–35.
12. Guddappa, S. I.; Hegde, R. M. Priority Aware Frequency Domain Polling Protocol in Cyber Physical Systems to Ejection of Malicious Node Attack. **2021,** *14* (3), 580–588.
13. Fang, W.; Zhang, W.; Yang, W.; Li, Z.; Gao, W.; Yang, Y. Trust Management-Based and Energy Efficient Hierarchical Routing Protocol in Wireless Sensor Networks. *Digit. Commun. Netw*. **2021**.
14. Mittal, N.; Singh, S.; Singh, U.; Salgotra, R. Trust-Aware Energy-Efficient Stable Clustering Approach Using Fuzzy Type-2 Modified Cuckoo Search Optimization Algorithm for Wireless Sensor Networks. *Wireless Netw.* **2010,** 1–24.
15. Sun, Z.; Wei, M.; Zhang, Z.; Qu, G. Secure Routing Protocol Based on Multi-objective Ant-Colony-Optimization for Wireless Sensor Networks. *Appl. Soft Comput.* **2019,** *77*, 366–375.
16. Qazi, R.; Qureshi, K. N.; Bashir, F.; Islam, N. U.; Iqbal, S.; Arshad, A. Security Protocol Using Elliptic Curve Cryptography Algorithm for Wireless Sensor Networks. *J. Ambient Intell. Human. Comput.* **2010,** 1–20.
17. Ahutu, O. R.; El-Ocla, H. Centralized Routing Protocol for Detecting Wormhole Attacks in Wireless Sensor Networks. *IEEE Access* **2020,** *8*, 63270–63282.
18. Shi, Q.; Qin, L.; Ding, Y.; Xie, B.; Zheng, J.; Song, L. Information-Aware Secure Routing in Wireless Sensor Networks. *Sensors* **2020,** *20* (1), 165.
19. Salari-Moghaddam, S.; Taheri, H.; Karimi, A. Trust Based Routing Algorithm to Improve Quality of Service in DSR Protocol. *Wireless Pers. Commun.* **2019,** *109* (1), 1–16.
20. Sun, Z.; Wei, M.; Zhang, Z.; Qu, G. Secure Routing Protocol Based on Multi-Objective Ant-Colony-Optimization for Wireless Sensor Networks. *Appl. Soft Comput.* **2019,** *77*, 366–375.

21. Yarde, P.; Srivastava, S.; Garg, K. Adaptive Immune-Inspired Energy-Efficient and High Coverage Cross-Layer Routing Protocol for Wireless Sensor Networks. *IET Commun.* **2020,** *14* (15), 2592–**2600**.
22. Sowmyashree, M. S.; Mala, C. S. Development of a Novel Protocol for Improvement of QoS in Wireless Sensor Networks: P-Rpeh. *Int. J. Recent Technol. Eng.* **2019,** *8* (3), 2277–3878.
23. Amin, S. U.; Hossain, M. S.; Muhammad, G.; Alhussein, M.; Rahman, M. A. Cognitive Smart Healthcare for Pathology Detection and Monitoring. *IEEE Access* **2019,** *7*, 10745–10753.
24. Olanas, et al. Smart Health: A Context Aware Health Paradigm Within Smart Cities. *IEEE Commun. Mag.* **2019,** *52* (8), 74–81.
25. Ayon, S. I.; Islam, M. M.; Hossain, M. R. Coronary Artery Heart Disease Prediction: A Comparative Study of Computational Intelligence Techniques. *IETE J. Res.* **2020.**
26. Islam Ayon, S.; Milon Islam, M. Diabetes Prediction: A Deep Learning Approach. *Int. J. Inf. Eng. Electron Bus.* **2019,** *11*, 21–7. https:// doi.org/10.5815/ijieeb.2019.02.03.
27. Islam, M. M.; Iqbal, H.; Haque, M. R.; Hasan, M. K. Prediction of Breast Cancer Using Support Vector Machine and K-Nearest Neighbors. In: *2017 IEEE Region 10 Humanitarian Technology Conference (R10-HTC)*; IEEE, **2017**; pp 226–229.
28. Hasan, M. K.; Islam, M. M.; Hashem, M. M. A. Mathematical Model Development to Detect Breast Cancer Using Multigene Genetic Programming. In: *2016 5th International Conference on Informatics, Electronics and Vision (ICIEV)*; IEEEl, 2016; pp 57457–9.
29. Haque, M. R.; Islam, M. M.; Iqbal, H.; Reza, M. S.; Hasan, M. K. Performance Evaluation of Random Forests and Artificial Neural Networks for the Classification of Liver Disorder. In: *2018 International Conference on Computer, Communication, Chemical, Material and Electronic Engineering (IC4ME2)*; IEEE, **2018**; pp 1–5.

CHAPTER 4

CPS Secured Hybridization Technique for Healthcare in Smart Cities

I. G. SARITHA[1], I. G. NAVEEN[2], and M. S. SOWMYASHREE[1]

[1]*Department of ETE, BMS Institute of Technology and Management, Bengaluru, Karnataka, India*

[2]*Department of ECE, Nitte Meenakshi Institute of Technology, Bengaluru, Karnataka, India*

ABSTRACT

Emerging wireless sensor network technology, cyber–physical systems (CPS), combines physical and computational capabilities. Due to the shared wireless channel and increasing system complexity, CPSs are vulnerable to malicious assaults. To address the requirements of high data rate in modern communications, the CPS are implemented in this work using an orthogonal frequency division multiplexing (OFDM) transmission device. The priority aware polling (PAP) is developed to find the required nodes from the pool of the nodes depending on its priority. In this, the priority of node is decided with data rate and throughput.

Additionally, the detected and prevented nodes, which are malicious, are used to reduce the loss of network packets. The performance analysis done for OFDM-used CPS in respect to packet delivery rate, throughput, packet loss ratio, and overhead. Using two widely used methods, like the trust aware wireless routing protocol (TAWP) and trust aware routing framework (TARF), the proposed hybridization method known as OFDM_PAP method, which combines the OFDM with PAP technology in

5G Green Communication Networks for Smart Cities. Devasis Pradhan, Rajeswari, Hla Myo Tun, Naw Khu Say Wah, & Thandar Oo (Eds.)

CPS is evaluated. It was found that the hybridized OFDM_PAP approach outperforms the TAWP and TARF protocols by a wide margin. The effectiveness of safe patient-centric healthcare applications using CPSs and various CPS architectures is also examined in this article. Compared to the current secured routing-based protocols, the proposed method offers better performance by decreasing the packet loss ratio and packet overhead, increasing throughput, and achieving maximum Packet delivery ratio. It also provides security by thwarting malicious nodes and extending the node's energy life. Such monitoring systems are crucial in both rural locations and smart cities to ensure that effective secured remote health-care monitoring is accomplished, reducing the need for patients to travel unnecessarily to a hospital and, as a result, lowering traffic, infection rates, and pollution.

4.1 INTRODUCTION

A fundamental concept of the fourth industrial revolution, the cyber–physical system (CPS), combines the physical and digital worlds in real time. It includes traits of hyperconnectivity, hyperautomation, and hyperintelligence.[15,22] The following four operations such as data acquisition, Analysis of data computation, biological object control, and formation of automatic data flow are the basis for the fusion of the material and cyber parts.[2,3] The CPS performs monitoring tasks using various autonomous and intelligent equipment, including robots, smart meters, actuators, sensors, controllers, servers, and gateways.[16] In CPS, several sensors are used to observe the physical amount. Different physical variables from the physical world, including humidity, temperature, rotational speed, and pressure, are monitored by the sensors of the CPS.[13] Modern vital infrastructures such as transportation, smart grids, water treatment, chemical plants and are thought to be built on top of the developed CPSs.[5,20]

Due to expanding communication networks used to monitor and manage physical systems, the CPS is more vulnerable to malicious attacks. Therefore, designing defenses against vicious assaults is challenging in network-controlled systems.[7,21] Additionally, the shared wireless medium communication which disrupts the operation of the physical devices which makes the CPS vulnerable to malicious cyberattacks.[4,9] When systems are subjected to insider and physical threats, traditional cyber security techniques such as authentication and encryption are rendered useless.[17] Complex algorithms are used in CPS for authentication, data integrity testing, and opponent detection for a significant amount of memory storage.[8] The primary aids of this study are as below

- The CPS utilizes a transmission mechanism using orthogonal frequency division multiplexing (OFDM) to fulfill modern communications' for necessary significant data rate.
- In OFDM-used CPS, the protocol priority aware polling (PAP) detects the higher-prior node, reducing node waiting time.
- In addition, network malicious nodes are identified, and are avoided to protect the transmission of data in the CPS.
- We examined the efficiency of safe patient-centric healthcare apps employing CPS with various CPS architectures because CPS has numerous applications in multiple sectors.

This proposed paper is structured generally as depicted as follows: Section 4.2 provides the connected research done on the most recent methods relating to the secure CPS. Section 4.3 describes the issue identified by the existing research and its resolution. Section 4.4 describes the proposed priority-aware frequency domain polling mechanism. Section 4.5 of the OFDM_PAP method findings and discussion is detailed. Section 4.6 discussed the application, Section 4.7 described the conclusion of the paper.

4.2 LITERATUTRE REVIEW

The study on the safety difficulties of CPS under untraceable assaults was presented by Zhao et al.[1] Using geometric control, it evaluates the design, implementation, and estimation of undetected attacks. The feed-forward structure was designed to achieve untraceable attacks in this CPS. The pole location of the assaulted system contributed to the development of the gain of the feed-forward/feedback structure. Here, three alternative approaches such as sensor attacks, actuator attacks, and coordinated actuators were measured to realize an undetectable attack. While analyzing the CPS undetectable attacks, this paper did not examine the packet delivery rate and delay.

The adaptive duty cycle control-based opportunistic routing (ADCCOR) approach were created by Xiang et al.[11] to achieve low power requirement with a greater response rate. The consumption of energy in wireless sensor networks played a significant role in the essential growth of this approach (WSN). When the node transmitter was about to send the data to the sink, the number of woken nodes rose in this approach. The network delay drastically decreased and dynamically altered the duty cycle of the sensor node. When the source node far away from the sink, the ADCCOR method's delay was lengthened. Additionally, this approach does not offer sufficient security to protect.

Gifty created the host-based probability intrusion detection, Bharathi and Krishnakumar,[6] utilizing maximum likelihood approximation and the Weibull approach. The compliance degree in this was used to identify the regular CPS nodes. The failure rate and dependability of the CPS and malicious mistakes were all examined using this intrusion detection technique. The strength of the system reaction was analyzed using the probabilities of false positives. Additionally, the system reliability was optimized using this probability value. However, due to the communication and noise issues in CPS, the calculation of compliance was unreliable. However, depending on the noise and communication faults, Different compliance levels were used to validate the node's state.

The stochastic modeling approach was presented by Shi et al.[18] to construct and address challenging attacks in CPS. The finite-state hidden Markov model (HMM) along with switching transition likelihood conditions was used to create the stochastic modeling outline. the Markov decision process manages the probability matrices of the switching transition. The attack approximation and joint state problem were articulated and resolved in the finite state, HMM, using the change in probability measure. Moreover, the attack and appropriate state were estimated using the marginal normalized conditional distributions.

Qureshi introduced the trust aware wireless routing protocol (TAWP), used to detach malicious attacks and recognize them from the network. To choose the best path among the trusted nodes, the TAWP was applied in four steps: Data collection, analysis of trust, route discovery, ranking, and route selection. In this TAWP, the trustworthiness of network nodes was confirmed using a trust analyzer. The trust data leads to finding the best path. At last, the information was sent and kept in the database. When the misbehavior happened during the data transmission, it impacted both the source and the route, which led to avoiding the transmission of data.

Chen et al.[14] introduced reliability and timeliness guaranteed opportunistic routing (RTGOR) to provide consistent data transmission of CPS. This protocol, which includes quantifiable transmission reliability and timing guarantees, was developed using the opportunistic routing approach. Since transmission delay and link delay in this protocol improved the transmission performance in CPS. This work does not consider the security besides malicious assaults that instigated the packet to be dropped across the network.

The CPS healthcare applications of patients were presented by Zhang et al.[23] They were based on big data analytics and the cloud. It gathers data using a standard, a service layer and a manageable data layer for parallel processing, and distributed storing. Utilizing big data and cloud technology can improve the efficiency of the healthcare system. The unified data collection, uniform APIs, and data-driven, cloud-enabled multisource heterogeneous healthcare systems were also presented in this study.

Jin Wang, et al. presented the newest architecture of CPS, dubbed cyber physical enhanced secured wireless sensor networks integrated cloud computing for u-life care), coupled with a healthcare monitoring application and decision support systems.[24] The main three parts of the proposed CPeSC3 architecture are communication, resource scheduling, and computation core and management core, which provides a thorough analysis of real-time scheduling, cloud computing, and security models. It also presents healthcare systems based on a 3-year practical test bed.

4.3 PROBLEM STATEMENT

This section describes the issues raised by the research that has already been done and the OFDM_PAP method used to resolve them. Communication failures and noise alter the degree of compliance that impacts the identification of the status of the node over CPS.[6] Due to misbehavior during the data transmission through CPS, the route is abandoned, and the data transmission is impacted.[19] Data broadcast over the network is affected by malicious attacks if no security plan exists.[14] A CPS with insufficient security results in packet loss when transferring data packets. The ADCCOR approach achieves increased latency when the relay node is too far from the sink node.

Solution:

The node with the highest priority is chosen using the PAP algorithm. Here, the data rate and throughput are used to determine the node's priority. When data is sensed from the OFDM transmitter to the receiver, this polling procedure aids in reducing the transmission latency. Additionally, each node's packet drop is examined to find network malicious nodes. Therefore, depending on the identification of malicious nodes, the injurious nodes are detected and avoided. By identifying rogue nodes, packet loss during data transmission is reduced.

4.4 OFDM_PAP METHOD

A secure OFDM-used CPS is created using the OFDM_PAP approach to meet contemporary communications' high data rate requirements. PAP is a protocol designed to identify nodes with higher priorities. In an OFDM-used CPS, the data rate and throughput determine the importance of the node. The nodes' packet drops are also identified to identify and avoid malicious nodes. As a result, the OFDM_PAP approach has higher throughput. Figure 4.1 displays the proposed method's flowchart.

4.4.1 DEVELOPMENT OF PRIORITY AWARE POLLING FOR CPS

OFDM-used CPS are initially thought of with K nodes and one sink node. Wideband OFDM communication channels are separated into 2S overlapping narrowband subcarriers. The divided subcarrier also has an equal bandwidth and constant noise power density. To overcome the extreme co-channel interference, this OFDM exclusively employs even subcarriers. Each frame comprises a D time slot in addition to the polling and transmission transit gaps to complete the OFDM broadcast. The broadcast packets are either rejected during communication, or the sink receives them in the preferred deadline (T frames).

PAP allows allocation for priority alert source (frequency) and node polling between the nodes, which results in reliable transmission. Additionally, as seen in Figure 4.2, the PAP generates the preamble phase, which the sink broadcasts to achieve time synchronization within the OFDM network. With reference to the response of frequency domain priority aware polling, the node priorities in the OFDM are gathered. Each node's data rate and throughput are used to determine the priorities. The sink resolves the uplink OFDM frequency problem to attain high data transmission based on the polling method.

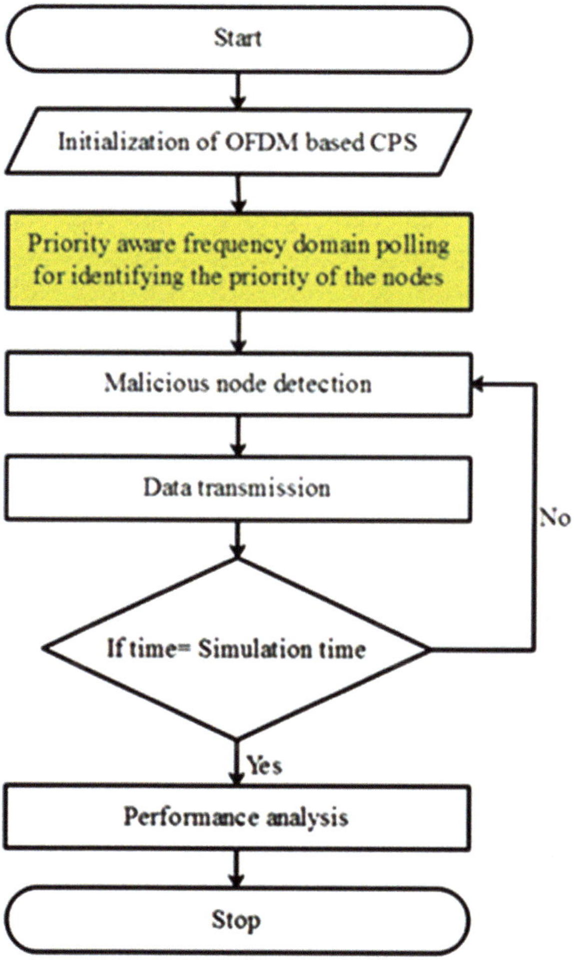

FIGURE 4.1 Flow diagram of OFDM_PAP method.

The aggregate acknowledgment from the destination node is used to allow the transmission of the data after the short interframe space (SIFS) period. Additionally, even though the scheduled node only delivers one packet per frame, it takes up to D slots.

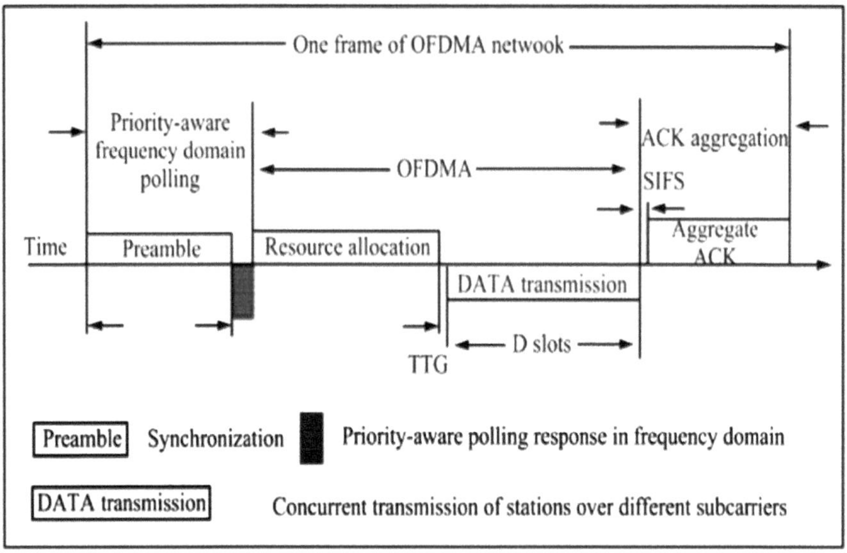

FIGURE 4.2 Architecture of PAP protocol.

During data transmission, the priority of the node is grouped into different tiers. In PAP, the node k (k = 1, 2, ... , K) and T_k [i.e., $s(k, 1)$, $s(k, 2)$, ... , $s(k, T_k)$] are allotted.

Here, the subcarriers determine the node's transmission priority, data rate, and throughput. The labels are defined as $1 = A(k, 1) A(k, 2) A(k, T_k) = T_k$, and the node priorities are represented as $A(k, i)$.

When packets are not conveyed effectively in the frame period, the node's priority will rise for the given time frame. The OFDM nodes communicate data packets to the sink node in each frame period based on the node priority.

Moreover, when harmful attacks require high priority during communication, the OFDM-used CPS's dependability is impacted. To prevent malicious attacks when transferring the packets, the CPS considers the nodes' packet drops during that time (i.e., more than 50% packet loss).

4.5 RESULTS AND DISCUSSION

This section details the findings and discussion of the OFDM_PAP approach. The Network Simulator 2.35 runs on a Windows 8 computer with

4GB RAM and an Intel Core i3 processor used to simulate and implement the suggested technique with an OFDM-used CPS. This approach makes use of the PAP protocol to obtain node polling. Additionally, the malicious nodes are identified with reference to the packet drops at each node. The OFDM-used CPS in this instance has 100 starting nodes spread throughout a 500–500 m^2 region. Table 4.1 displays the specifications taken into account for this OFDM_PAP technique.

TABLE 4.1 PAP Specifications.

Parameter	Value
Total number of Node	100
Total_area	$500 \times 500m^2$
Transmission channel mode	Wireless
Method	Two ray propagation
Antenna used	Omni
Queue/length	Priority/200
Protocol type	MAC_802_11

4.5.1 PERFORMANCE ANALYSIS

Packet delivery ratio (PDR), PLR, Throughput, and overhead, are used to analyze the OFDM_PAP method's performance. By changing the number of malicious nodes from 1 to 5, the effectiveness of this technique is assessed and contrasted with the TAWP and TARF.

4.5.2 PACKET DELIVERY RATIO

It is the proportion of the total number of transmitted packets to that are acknowledged by the sink (7).

$$PDR = \frac{\sum_{i=1}^{n} Y_i}{\sum_{i=1}^{n} X_i} \times 100 \quad (4.1)$$

where the total amount of packets received at the sink is signified as Y; the total amount of transmitted packets is denoted as X, and the number of transmitted nodes at the source-used CPS is denoted as i.

TABLE 4.2 PDR Comparison.

Malicious nodes	TAWP	TARF	OFDM_ PAP
0	100.0	100.0	100.0
1	87.1	85.1	99.75
2	83.1	77.1	99.51
3	80.1	60.1	99.31
4	73.1	50.1	99.61
5	60.1	40.1	99.41

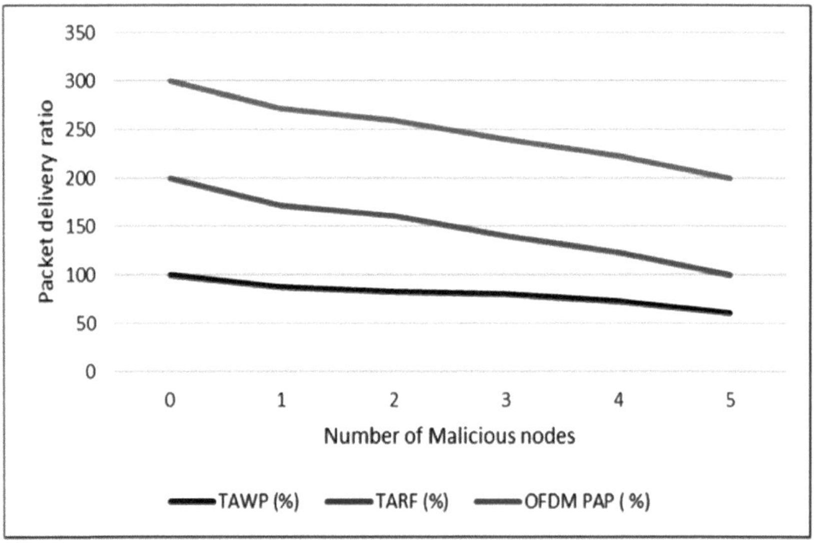

FIGURE 4.3 Graph of PDR comparison.

With reference to the limitations of malicious nodes, the PDR of the OFDM_PAP technique is raised by 99.41% in comparison to TARF and TAWP, where the PDR is 40.1 and 60.1%, correspondingly, as in Table 4.2 and Figure 4.3.

4.5.3 PACKET LOSS RATIO

It is a ratio of the total amount of transmitted data packets to the total amount of data packets lost. The PLR expression for the same is shown in eq. (4.2).

$$PLR = \frac{\sum_{i=1}^{n} X_i - Y_i}{\sum_{i=1}^{n} X_i} \times 100\% \tag{4.2}$$

TABLE 4.3 PLR Comparison.

Malicious nodes	%TAWP	%OFDM_PAP
1	11.41	0.071
2	10.56	0.11
3	12.30	0.15
4	12.6	0.081
5	10.59	0.120

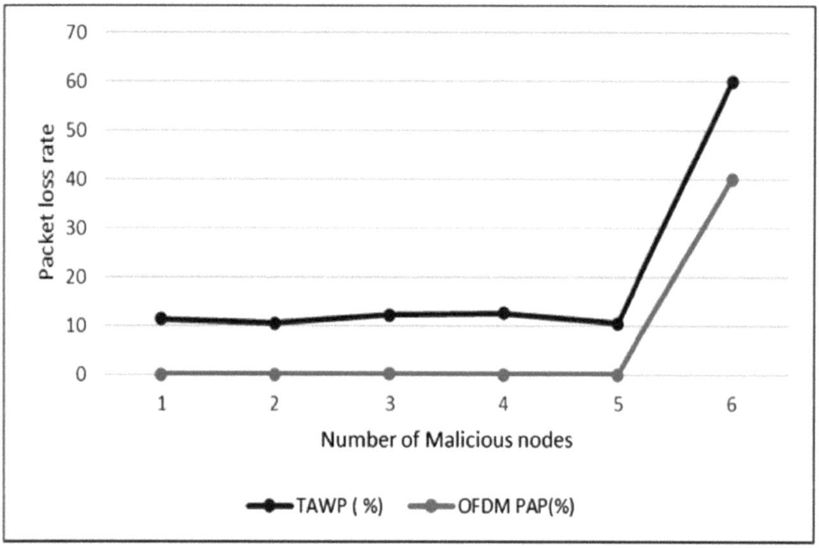

FIGURE 4.4 Graph of PLR comparison.

The PLR of this approach is reduced by 0.120% for five malicious nodes in relation to the TAWP which is 10.59% because of the network nodes' misbehavior as in Figure 4.4 and Table 4.3.

4.5.4 OVERHEAD

It is the proportion of the number of control packets formed by the network nodes as stated in the eq (4.3).

$$Overhead = \sum_{j=1}^{n} R_j \quad (4.3)$$

where the number of nodes that formed the control packets is denoted as j and the control packets are denoted as R.

TABLE 4.4 Overhead Comparison.

Malicious nodes	TAWP	TARF	OFDM_PAP
1	1.1	1.6	1.4
2	1.5	2.1	1.5
3	1.91	3.1	1.8
4	2.1	4.1	2.21
5	2.91	5.61	2.61

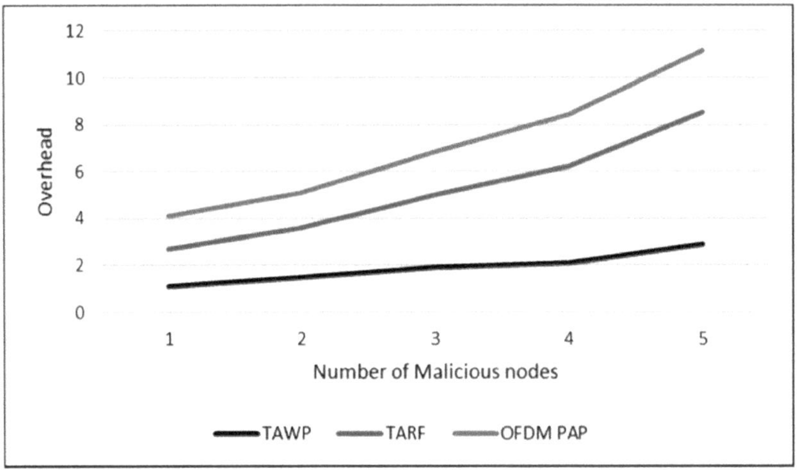

FIGURE 4.5 Graph of overhead comparison.

The overhead of this method is 2.61 control packets for five malicious nodes which is low because of priority in comparison to TARF and TAWP as shown in Figure 4.5 and Table 4.4.

4.5.5 THROUGHPUT

It is defined as the amount of data packets effectively received by the sink in the total time interval that is stated in eq. (4.4).

$$Throughput = \frac{Y_i}{TI} \tag{4.4}$$

Where, T states the time period and Yi specifies the data packets acknowledged at the sink.

TABLE 4.5 Throughput Comparison.

Malicious nodes	TAWP [bps]	OFDM_PAP (bps)
1	76.461	162.191
2	83.881	163.951
3	75.391	161.891
4	70.291	162.141
5	69.381	160.961

It achieves maximum throughput from this method which is 160.961 for five malicious nodes related to the TAWP as in Figure 4.6 and Table 4.5.

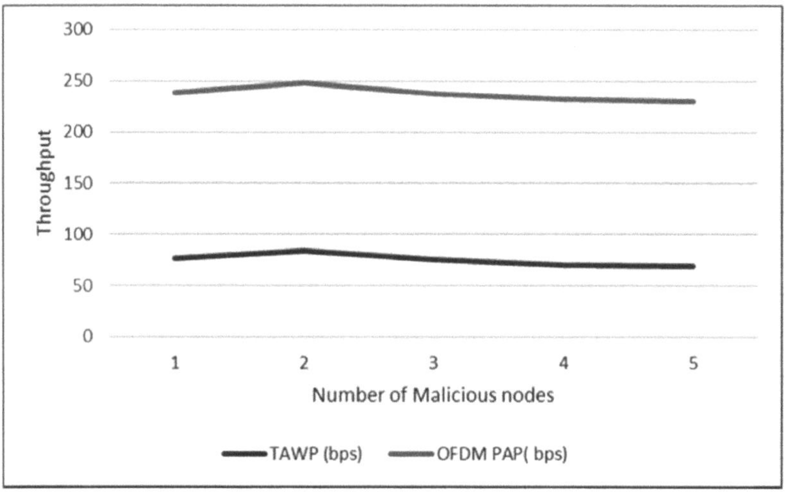

FIGURE 4.6 Graph of throughput comparison.

4.6 CPS HEALTH CARE ARCHITECTURE

The CPS healthcare architecture is shown in Figure 4.7—the necessity for telemedicine is growing significantly due to the lack of physicians in rural and outlying areas and pandemic scenarios such as COVID-19. With the Internet, doctors, and patients can communicate with one another over the phone or through computer systems. Data security is crucial since patients will communicate their issues and receive medical assistance across the network. Therefore, physical and cyber implications must be considered while designing CPS for healthcare. The proposed Secured hybridization technique for CPS is utilized for telemedicine-based healthcare monitoring.

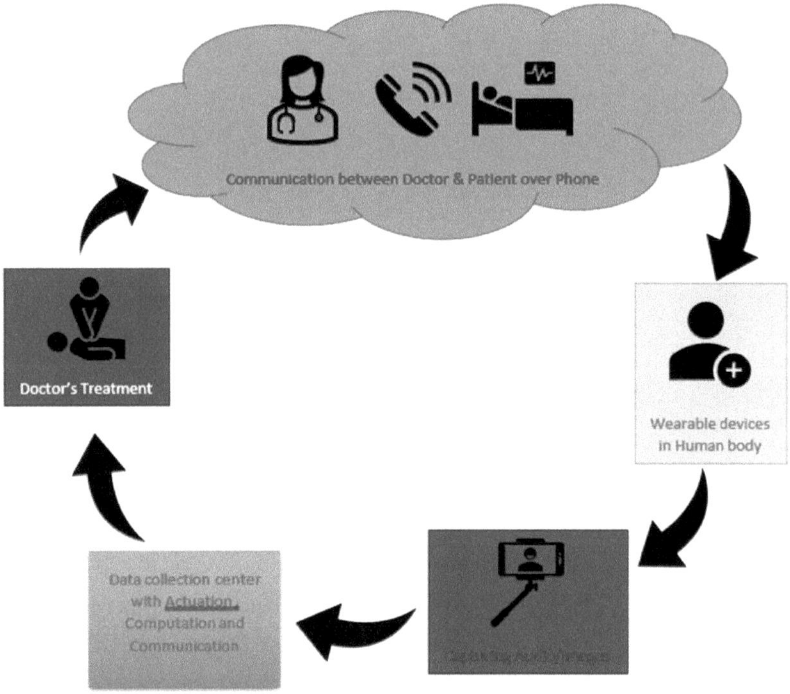

FIGURE 4.7 CPS Healthcare architecture.

While sitting at home, the patient's data can be sensed and gathered by sensors attached to the human body or the walls of the home environment. This audiovisual-based approach collects images from cameras, extracts the background to know the movement of objects, and identifies various activities like sleeping, sitting, walking, running, and jumping, among others. Using an activity-based access control system with image-based authentication improves user access flexibility and security. Modules for computing and communication can both use the security unit.

4.7 CONCLUSION

To provide efficient communication and efficient polling, priority aware polling is created in this research using an OFDM-used CPS. The data rate and throughput are used to determine the nodes' priority. The created PAP protocol is used to reduce the time packets must wait while being transmitted, and network malicious nodes are found based on the drop of each packet at the node. Therefore, the OFDM_PAP approach is employed to achieve a high data rate compared to the existing methods to fulfill the modern communication systems.

Compared to TAWP and TARF, the suggested technique performs better for five malicious nodes in terms of PLR of 0.120%, Overhead of 2.61, Throughput of 160.961 bps and PDR 99.41% are achieved efficiently. The effectiveness of healthcare systems using various types of CPS design is also evaluated for social necessity. In the future, polling overflow in CPS can be eliminated using frequency interleaved polling and can develop healthcare designs further to provide greater performance.

KEYWORDS

- **malicious nodes**
- **cyber-physical systems**
- **priority aware polling**
- **OFDM**

REFERENCES

1. Zhao, Z.; Yang, Y.; Li, Y.; Liu, R. Security Analysis for Cyber-Physical Systems Under Undetectable Attacks: A Geometric Approach. *Int. J. Robust Nonlinear Control* **2020,** *30* (11), 4359–4370.

2. Lyu, X.; Ding, Y.; Yang, S.H. Bayesian Network Based C2P Risk Assessment for Cyber-Physical Systems. *IEEE Access* **2020,** *8*, 88506–88517.
3. Yuan, Y.; Mo, Y. Security for Cyber-Physical Systems: Secure Control Against Known-Plaintext Attack. *Sci. China Technol. Sci.* **2020,** 1–10.
4. Ning, X.; Jiang, J. In the Mind of an Insider Attacker on Cyber-Physical Systems and How Not Being Fooled. *IET Cyber-Physical Syst.: Theory App.* **2020,** *5* (2), 153–161.
5. Zhang, Y.; Yağan, O. Robustness of Interdependent Cyber-Physical Systems Against Cascading Failures. *IEEE Trans. Autom. Control* **2019,** 65 (2), 711–726.
6. Gifty, R.; Bharathi, R.; Krishnakumar, P. Privacy and Security of Big Data in Cyber Physical Systems Using Weibull Distribution-Based Intrusion Detection. *Neural Comput. App.* **2019,** *31* (1), 23–34.
7. Lima, P. M.; Alves, M. V. S.; Carvalho, L. K.; Moreira, M. V. Security Against Communication Network Attacks of Cyber-Physical Systems. *J. Control, Autom. Electr. Syst.* **2019,** *30* (1), 125–135.
8. Mili, S.; Nguyen, N.; Chelouah, R. Transformation-Based Approach to Security Verification for Cyber-Physical Systems. *IEEE Syst. J.* **2019,** *13* (4), 3989–4000.
9. Kim, S.; Won, Y.; Park, I. H.; Eun, Y.; Park, K. J. Cyber-Physical Vulnerability Analysis of Communication-Based Train Control. *IEEE Internet Things J.* **2019,** 6 (4), 6353–6362.
10. Sowmyashree, M. S.; Mala, C. S. Development of a Novel Protocol for Improvement of Qos in Wireless Sensor Networks: P-Rpeh. *Int. J. Recent Technol. Eng.* **2019,** *8* (3).
11. Xiang, X.; Liu, W.; Liu, A.; Xiong, N. N.; Zeng, Z.; Cai, Z. Adaptive Duty Cycle Control–Based Opportunistic Routing Scheme to Reduce Delay in Cyber Physical Systems. *Int. J.Distrib. Sens. Netw.* **2019,** *15* (4), 1550147719841870.
12. Fu, R.; Huang, X.; Xue, Y.; Wu, Y.; Tang, Y.; Yue, D. Security Assessment for Cyber Physical Distribution Power System Under Intrusion Attacks. *IEEE Access*, 7, 75615–75628.
13. Orojloo, H.; Azgomi, M. A. A Stochastic Game Model for Evaluating the Impacts of Security Attacks Against Cyber-Physical Systems. *J. Netw. Syst. Manag.* **2018,** *26* (4), 929–965.
14. Chen, A.; Li, X.; Ni, X.; Luo, G. RTGOR: Reliability and Timeliness Guaranteed Opportunistic Routing in Wireless Sensor Networks. EURASIP *J. Wireless Commun. Netw.* **2018,** *2018* (1), 86.
15. Lee, B. M.; Yang, H. Massive MIMO for Industrial Internet of Things in Cyber-Physical Systems. *IEEE Trans. Ind. Inf.* **2017,** *14* (6), 2641–2652.
16. Alcaraz, C.; Lopez, J. A Cyber-Physical Systems-Based Checkpoint Model for Structural Controllability. *IEEE Syst. J.* **2017,** *12* (4), 3543–3554.
17. Zheng, B.; Deng, P.; Anguluri, R.; Zhu, Q.; Pasqualetti, F. Cross-Layer Codesign for Secure Cyber-Physical Systems. *IEEE Trans. Comput.-Aid. Design Integr. Circ. Syst.* **2016,** *35* (5), 699–711.
18. Shi, D.; Elliott, R. J.; Chen, T. On Finite-State Stochastic Modeling and Secure Estimation of Cyber-Physical Systems. *IEEE Trans. Autom. Control* **2016,** *62* (1), 65–80.
19. Qureshi, N. Malicious Node Detection Through Trust Aware Routing in Wireless Sensor Networks. *J. Theoret. Appl. Inf. Technol.* **2015,** *74* (1).

20. Huang, S.; Zhou, C. J.; Yang, S. H.; Qin, Y. Q. Cyber-Physical System Security for Networked Industrial Processes. *Int. J. Autom. Comput.* **2015,** *12* (6), 567–578.
21. Hahn, A.; Thomas, R. K.; Lozano, I.; Cardenas, A. A Multi-Layered and Kill-Chain Based Security Analysis Framework for Cyber-Physical Systems. *Int. J. Crit. Infrastruct. Protect.* **2015,** *11*, 39–50.
22. Wan, K.; Alagar, V. Context-Aware Security Solutions for Cyber-Physical Systems. *Mob. Netw. App.* **2014,** *19* (2), 212–226.
23. Zhang, Y., et al. Health-CPS: Healthcare Cyber-Physical System Assisted by Cloud and Big Data. *IEEE Syst. J.* **2015**.
24. Wang, J.; Abid, H.; Lee, S.; Shu, L., Xia, F. A Secured Health Care Application Architecture for Cyber-Physical Systems. *Control Eng. Appl. Inf.* **2011.**

CHAPTER 5

Flexible Diodes: Principles, Concepts, and Applications Toward 5G Green Communication

ARUN AGARWAL[1], VARUN SHARMA[2], GOURAV MISRA[2], and SAURABH NARENDRA MEHTA[3]

[1]*Department of ECE, ITER, Siksha 'O' Anusandhan Deemed to be University, Bhubaneswar, Odisha, India*

[2]*School of Electronic Engineering, Dublin City University, Dublin, Ireland*

[3]*Department of ECE, Vidyalankar Institute of Technology, Mumbai, Maharashtra, India*

ABSTRACT

Different types of flexible diodes are fabricated to get gains in wide number of industries. Implementation of flexible electronics in light-emitting diode (LED) has increased its durability, and flexibility and in some cases made it transparent in the visible spectrum. Making use of organic dopants in the emission layer of LED and getting a white spectrum which can later be spliced up into different colors is discussed. Phosphorescent emitters along with polyethylene naphthalate substrates can be used to produce a flexible white organic light-emitting diode (FWOLED). Manufacturing of radio frequency diodes, transistors, and switches using silicon and germanium nanomembranes used for advancements in near-field and far-field communications. Using graphene to form graphene field effect transistors for radio frequency applications was also studied. Transparent

5G Green Communication Networks for Smart Cities. Devasis Pradhan, Rajeswari, Hla Myo Tun, Naw Khu Say Wah, & Thandar Oo (Eds.)

flexible high-voltage diodes used for better power management in wearable devices with the help of triboelectric nanogenerator (TENG) generators, zinc oxide channels, and indium tin oxide material electrodes to make the device flexible, transparent, and also perform high-voltage rectification being discussed. Fabrication methods such as roll-to-roll printing for printing on flexible substrates are also popular among flexible electronics manufacturing.

5.1 INTRODUCTION

Developments in rigid semiconductor devices produced amazing results in the field of photonics, high-voltage devices, transistors, and diodes. Following Moore's law, the number of transistors in a chip increased gradually every year with the latest fabrication techniques and complementary metal–oxide semiconductor (CMOS) technology developments in materials such as silicon, germanium, and oxide dielectric material, such as hafnium oxide were used. Along with these improvements, the next step was making these materials flexible. The aim behind production was to make these devices lightweight, thinner, and nonbreakable which can later on be implemented in a number of applications.

Flexible electronics along with the organic counterpart led to an even further boost in device properties ranging from optics, radio frequency, and flexibility of the device. Manufacturing of the organic industry was similar to the printing device leading to the printing of flexible substrates which were the fundamentals of flexible organic devices. Flexible electronics were divided into four different types of applications ranging from sensors, batteries, and solar cells, electronic displays, and circuits. The list of applications produced is very large in number as most of them are commercialized and used on a daily basis. This paper will go through some of the flexible devices such as FWOLED, flexible diodes for radio frequency, and high-voltage flexible transparent diodes. There are many other applications of flexible devices as well such as flexible speakers, flexible pressure and touch sensors, flexible lightning modules, flexible photovoltaics, and flexible displays.[1] Manufacturing of such devices has many issues to overcome such as substrates facing temperature, mechanical deformities and fabrication, quality of organic material fabrication, conductivity, and work functions are also some of the factors to overcome. The major thing that can affect development is price and fabrication process. The merger of flexible and organic electronics has led to developments in the medical sector such as recording and stimulating neural activity to detect various neurological disorders using polymer electrodes.[2] Similarly, the properties, application, and fabrication process of such flexible electronics will be discussed in this paper related to the area of light, radio frequency, and high-voltage transparent diodes.

5.1.1 FWOLED

Since the invention of the light bulb to an era where there has been a new form of pollution in form of light known as light pollution. There were number of developments in the way light was emitted in terms of color temperature and voltage efficiency ranging from incandescent, and high intensity to LED. Evolution in organic semiconductors where it is possible to print circuits with organic semiconductors dissolved in a solvent to produce organic semiconductor ink.[3] Further research led to an organic semiconductor electroluminescent emits light by applying some voltage across electrodes. It gave rise to organic light-emitting diode (OLED) which had amazing properties such as quantum efficiency, high luminance, power efficiency, and more life compared to LED.[4] There were two methods to get the color display output, the first being the selective deposition method where RGB color was patterned individually. The second one is combining the micropatterned color filters with the WOLED to obtain the three colors.[3] Color rendering index CRI is the factor used to measure the level of reproduction close to its natural level with a specified illumination condition.[5] Figure 5.1 depicts the different CRI ratios from halogen and incandescent bulbs which is 100 to light sources such as Sodium and Mercury where the CRI gradually declines below 50. A light source which has a CRI value of more than 80 Ra is considered a good light source and the color temperature should also be in the range of 3000K to 6000K.[3]

Research on OLED led to the emergence of other optoelectronics fields. However, one region that was widely researched is the fabrication of OLED. OLED was fabricated on a plastic or glass substrate and this method was used in various commercial devices such as TVs, mobile phones, etc. Substrates later evolved to be malleable to all mechanical deformities by using a flexible substrate over which the OLED was fabricated resulting in a FWOLED.[5]

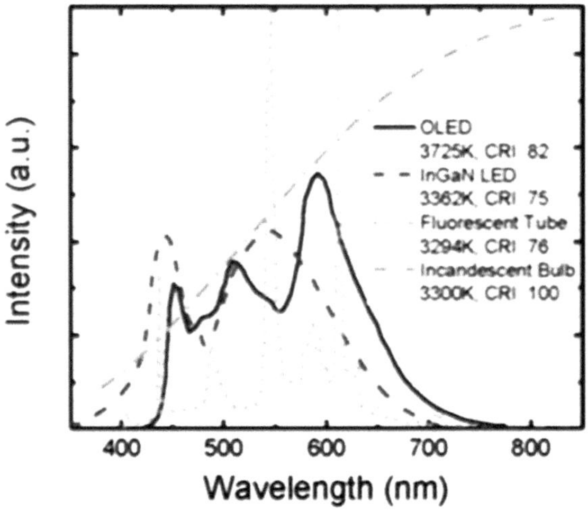

FIGURE 5.1 CRI for different sources of light.

The FWOLED differs from the LED in the bottom substrate where a flexible substrate is used in the case of FWOLED. Preference for OLED in the commercial market is slowly rising with its high-quality display, color rendering index, luminance, and color stability.

The basic structure of an OLED with a glass substrate consists of the following materials. As the photons are emitted in the downward direction toward the glass substrate the lowermost material is a glass substrate. The material above the substrate is an anode made up of Indium tin oxide which is responsible for hole injection in the device. Over the anode, there is a hole transport layer that consists of organic materials. An organic material is good at conducting electrons or holes. Above this hole transparent layer, there is a host layer which is also known as the emissive layer and consists of luminescent dopants with a wide energy band gap. Hosts that are bipolar help in a higher rate of electron–hole recombination process resulting in greater returns of luminesce.[6] Over the host layer, there is an electron transport layer which also consists of organic materials. On the top of this layered structure, there is a cathode made up of LiF and Al which is responsible for electron injection into the device. The difference between an OLED and an LED is the presence of organic materials used for photon emission between the cathode and anode.[5]

The working principle of luminescence can be described with relation to the *p-n* junction diode. When a certain voltage is applied to bias the diode, which can be around 2–3 volts, electron–hole pairs are formed in the host region due to injected charge carriers. This excitation is created between the organic layers. Due to the presence of luminescent dopants, there is emission of photons from the host layer.[5]

With this OLED structure, it can be further bifurcated into white OLEDs. There are ways to generate this white light from OLED sources. First, white light is generated by mixing primary colors such as red, green, and blue by using multiple emitters. In the second method, shorter wavelengths emissions from OLED are mixed together to excite phosphors which incur as white light to the human eye.[7] The performance of WOLED is based on how the charge and exciton distribution are managed to yield the best triplet state which is the three allowed values of the spin component. Phosphorescent emitters are used in this case to yield the maximum output from Triplet State.[4] Such low molecular weight luminescent materials aid in increasing the efficiency of OLED.[7]

Use of flexible substrates in WOLED can result in FWOLED. The emission pattern of these emitters can be divided into different configuration such as top emitting, bottom emitting, transparent OLED, and tandem OLED.[8] The structure of FWOLED can be described as follows from bottom to top. The base layer consist of substrate which is flexible followed by an electrode 1. This electrode 1 is patterned as per the requirement or emission which can be opaque for top-emitting FWOLED and transparent for the bottom-emitting diode. This layer is followed by the organic first injection and transport layer and the second injection and transport layer, which encapsulates the emitter layer between them.[4] The emitter layer can be related to the host dopant layer of OLED devices. Electrode 2 is on top of this structure above the charge injection layer CIL the property of this layer is varied similar to electrode 1 wherein, for bottom-emitting diodes it is opaque and transparent for top-emitting diodes. The major difference between WOLED and FWOLED is the flexible nature of CIL, CTL, and electrodes including the substrate.

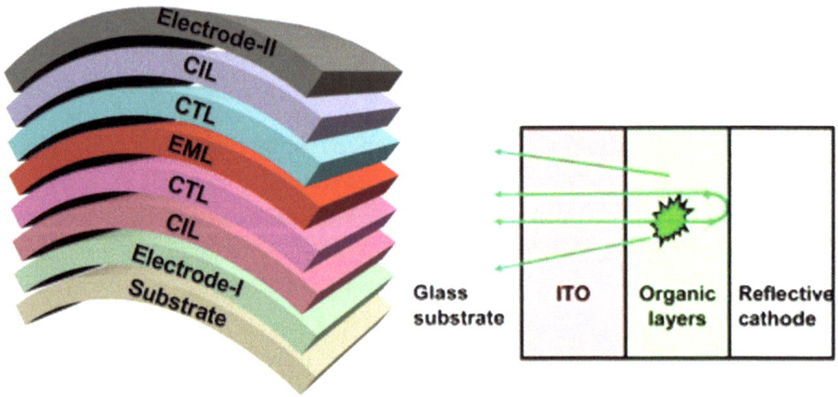

FIGURE 5.2 (a) Structure of FWOLED[4] and (b) bottom emitting substrate.

Source: (a) Reprinted from Ref. [4]. Copyright © 2019 by the authors. Licensee MDPI, Basel, Switzerland. http://creativecommons.org/licenses/by/4.0/. (b) Reprinted from Ref. [8]. Copyright © 2010 by the authors; licensee Molecular Diversity Preservation International, Basel, Switzerland. http://creativecommons.org/licenses/by/3.0/.

5.1.2 FLEXIBL DIODE FOR RADIO FREQUENCY

The need of wireless technology is growing rapidly and has occupied all the industries starting from healthcare, shopping centers, manufacturing industries, banking, and automobiles. This development was taken to the next level with the help of the Internet of Things, machine learning, transparent, flexible, and cheaper forms of microchip fabrication. Key fundamentals of these latest inventions lie in the fundamental electronics development. It also started in Feb 2005, when the first transfer-fitted flexible Si DC MOSFET was developed till the point where high-speed flexible electronics were researched to work in high frequency (microwave range).[10,11] The ultimate aim is to merge microwave engineering with flexible materials to obtain the transfer of energy/signals.[10]

One of the most widely used applications that works on the principle of data transmission with the help of radio frequency is radio frequency identification (RFID) tags. These tags are widely accepted and used in various day-to-day applications such as hospital patient data records, vehicular data collection, Key tags in hotels and residential security, and many more. The basic operation of this RFID can be explained with the help of an RFID reader/transmitter along with an RFID tag. The tag comprises many microelectronic components such as a rectifier, an antenna, diodes, and logic circuitry. The working of this transmission starts with the emitter sending an AC signal. This AC signal is picked up by the antenna of RFID tag. The signal is then rectified in the tag to produce a rectified direct current which can be used to power the logic circuit to generate the unique code stored in the memory of this tag. The code sequence can then be obtained using the load modulator which works in on and off state utilizing voltage generated from the rectifier. Depending on RFID tags are categorized into two types, passive tags where there the circuit of the tag relies on the signal from the emitter and second active tags where the voltage required from the circuit is obtained using a battery.[12]

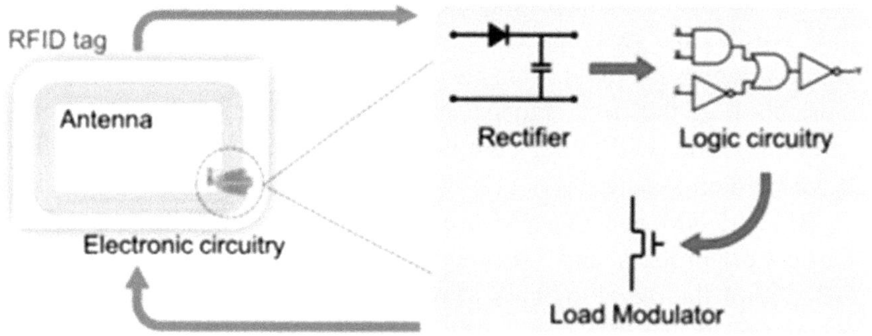

FIGURE 5.3 RFID tag running at 13.56 MHz.

Source: Reprinted from Ref.[12]. https://creativecommons.org/licenses/by/3.0/

The development of this technology started with MHz and slowly performed in GHz range, lower frequency also resulted in less coverage area. The close-distance technology which was also derived and was successful in communicating data is near-field communication. It is used in mobile phones to transmit/receive data with other RFID cards/tags and also with other mobiles to transmit data.

Developments in this field to make such components flexible and to support high-frequency range in microwave range with high mobility one single-crystal nanomembranes and producing thin film transistors, PIN diodes, and single pole single throw (SPST) switches.[11] To produce such flexible transistors with high mobility carriers using selective etching and transfer printing. The process of fabrication begins with silicon on an insulator substrate. The top layer of this substrate consists of silicon which is etched using reactive ion etching or lithography process to get the required shape and size. The shapes can be long ribbons or small holes. With the help of concentrated hydrofluoride HF solution to etch the exposed buried oxide. Once the etching is done, the silicon nanomembrane which is transferred to a flexible substrate and is held together using an adhesive epoxy coating. It slowly gets bonded to form a single Silicon NM layer.[11]

Radio frequency diodes consists of inductors, capacitors, switches, and diodes and to be termed as a flexible radio frequency all these electronic components should be manufactured similarly. These flexible switches and diodes can be manufactured using Si NM and Ge NM materials. The procedure to produce such flexible switches is similar to the one discussed with *p*-type doping on the type using ion implantation. For flexible switches, the gate bridges are also removed from it the material which makes it more flexible. The process being similar to the fabrication of thin film transistors helps manufacture flexible switches and compatible with other complementary metal–oxide semiconductor (CMOS) components.[11]

Both Si NM and Ge NM are good material both having different fabrication process and are capable of working in the range of 20 GHz. With the developments in the semiconductor industry there was researches in two materials gallium nitride (GaN) and gallium arsenide (GaAs). Using a high electron mobility transistor with GaN fabricated on plastic substrates showed good response for power amplification and high-frequency responses. For the other material, GaAs used as a heterojunction bipolar transistor also showed good frequency responses. The research was conducted to fabricate GaAs on biodegradable cellulose substrates.[10] These materials were further developed to inculcate other electrical components and were widely used as high-frequency actives in a microwave communication system. Whereas in passive systems, the developments of flexible passive were slow as they involved the manufacturing of flexible capacitors and inductors.[10]

Graphene field effect transistors were also used in radio frequency flexible diode sections. Graphene is a better option in terms of strain limits compared to Si film semiconductors. Fabrication of GFETS is done with the help of a chemical vapor deposition process. Graphene is deposited on the top of the Hafnium dioxide layer and the channel is connected between

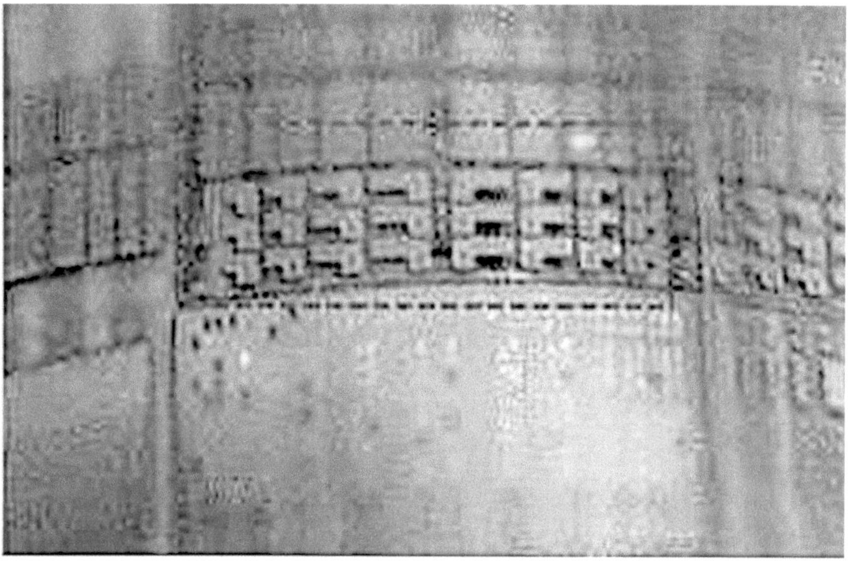

FIGURE 5.4 Flexible SPST switch on a plastic.

the source and drain.[13] It was observed that to increase the mechanical deformation property of this material, an electrode with a good resistance to strain can increase its flexibility. A case of electrode breaking at the junction and creating of scenario of an open circuit was observed when the material was deformed above its strain limits. Another factor that affected the performance of radio frequency is the thermal management of the substrates.[13]

5.1.3 FLEXIBLE TRANSPARENT HIGH-VOLTAGES DIODES

Recent developments in flexible diodes have led to a wide variety of applications ranging from displays, identification tracking cards, and vehicle toll cards to handheld devices such as bendable phones, wearable devices, etc. Basic need of all these devices being very high mechanical deformation sustainability, power performance, backward compatibility or a widely used fabrication method. As device sizes are reduced so the complexity of power management in these devices are also increasing. Wearable devices are slowly attracting customer's attention worldwide as their applications are not limited. These devices can perform task such as heart rate monitoring, altitude monitoring, humidity monitoring, and fitness exercise calculations deriving from motions, video, audio, and even telecommunication that too all packed in a small wearable or handheld device. To perform so many tasks given its considerably small size, managing power is one of the key factors in developing such devices efficiently. This energy efficiency in wearable devices was optimized by using a triboelectric nanogenerator (TENG).[14]

This material is used in self-powered wearable device where the electricity is generated using the ambient motions and is very lightweight and compact which makes it perfect for wearable devices. Flexible rectifiers are generally used along with TENG to convert the high alternating current to the low direct current voltage. Power management of these TENGs is of major importance as it will affect the performance and battery life of the device.[15] Fabrication of flexible devices takes place at temperatures below 200°C whereas high-voltage diodes such as silicon carbide, GaN requires high temperatures to manufacture. Zinc oxide is most preferred inorganic material in low temperature processing and also due to its high electric performance and transparency ratio.[14]

Fabrication of such wearable electronics is carried out using the prevalent lithography techniques and methods used in film synthesis. This method is also used for production of other flexible devices which makes it compatible with other flexible CMOS electronics. The process starts with polyethylene naphthalate substrate widely used for flexible devices over which Cr-anode is placed at the bottom. Dielectric Al_2O_3 is deposited using the atomic layer deposition process. Conductivity of the channel is varied by the Cr anode and Al_2O_3 to form a metal insulator semiconductor device.[16] ZnO layer is later deposited which forms the channel for this device. Cathode is placed on the edge of the circle while the anode is located in the center. The structure of this placed in such practice that it forms an ohmic contact with the ZnO channel.[14]

The channel of this device which is made up of ZnO can be divided into high electron concentration and low electron concentration regions. When a positive bias is applied, all the electrons are collected in the gate region which increases the conductivity of it on the other hand it becomes highly resistive when the negative bias is applied which leads to depletion of the electrons in the gate region.[14] Being a flexible device, it should pose high mechanical deformities as the device can be flexed for a large number of times. The effect of this bending and flexing can be demonstrated via the *I-V* graph. The graph demonstrated that after 10,000 bends there was a slight voltage drop. Bending tests are performed on this device to measure the potential of these flexible devices. It was also observed that after a certain number of bends the electrode cracks up and when this substrate is again brought to a flat surface the cracks united together which slowly led to the recovery of the current.[14]

The most popular material used to make these flexible diodes transparent is indium tin oxide as a transparent substrate. This material possesses excellent optical transparency, good electrical conductivity, and work function for hole injection being a metal oxide it also shows good durability.[17,18]. Due its rigid nature there are high chances of ITO material to develop cracks which can affect the conductivity. To make it compatible with flexible devices these ITO electrodes are structured in a mesh. With the help of a flexible film, the structured mesh pattern is imposed on it. Using this mesh pattern helps to increase the strain resistivity of the whole material and also avoids loss of conductivity if there are cracks by suppressing it.[18] These materials can be fabricated using wet etching and photolithography process. Development in every single minute component of semiconductor has led to a product which can be transparent, flexible and also capable of working with high voltage. Diodes of such types are improving rapidly in the field of wearable electronics.

5.2 CONCLUSION

It was found that there are still issues and problems with flexible and organic electronics. While certain factors increase the efficiency while others affect the manufacturing process. The flexible FWOLED device still are far from commercialization and research in the field of TADM emitters is broadly open and building up. Other factors related to flexible devices are the lifetime of the device which still needs further enhancements. Research in radio frequency devices is to obtain far-field communications in microwave frequencies. Single crystalline silicon rectifiers along with silicon and germanium nonmembrane play a key role in fabrication and performance. Large surface area deposition has the worst radio frequency efficiency. High-voltage diodes are showing properties that make them compatible with TENG and further the energy management in wearable devices.

KEYWORDS

- **LED**
- **spectrum**
- **radio frequency**
- **silicon**
- **graphene**

REFERENCES

1. Tsai, C.-C. Recent Development in Flexible Electronics. In: *16th Opto-Electronics and Communications Conference*; **2011**; pp 370–371.
2. Cicoira, F. Flexible, Strechable and Healable Electronics. In: *2018 International Flexible Electronics Technology Conference (IFETC)*; **2018**; pp 1–1.
3. Zyung, T. et al. Flexible Organic LED and Organic Thin-Film Transistor. *Proc. IEEE* **2005,** *93* (7), 1265–1272.
4. Luo, D.; Chen, Q.; Liu, B.; Qiu, Y. Emergence of Flexible White Organic Light-Emitting Diodes.*Polymers* **2019,** *11* (2).
5. Chang, Y.-L.; Lu, Z.-H. White Organic Light-Emitting Diodes for Solid-State Lighting. *J. Disp. Technol.* **2013,** *9* (6), 459–468.
6. Navamani, K.; Samanta, P. K.; Pati, S. K. Theoretical Modeling of Charge Transport in Triphenylamine–Benzimidazole Based Organic Solids for Their Application As Host-Materials in Phosphorescent OLEDs. *RSC Adv.* **2018,** *8* (52), 30021–30039.
7. Destruel, P.; Ablart, G.; Jolinat, P.; Seguy, I.; Farenc, J. White Organic Light-Emitting Diodes (WOLEDs).In: *Conference Record of the 2006 IEEE Industry Applications Conference Forty-First IAS Annual Meeting*, Vol. 2; **2006**; pp 694–697.
8. Wei, M.-K.; Lin, C.-W.; Yang, C.-C.; Kiang, Y.-W.; Lee, J.-H.; Lin, H.-Y. Lin. Emission Characteristics of Organic Light-Emitting Diodes and Organic Thin-Films with Planar and Corrugated Structures. *Int. J. Mol. Sci.* **2010,** *11* (4), 1527.
9. Aleksandrova, M. Specifics and Challenges to Flexible Organic Light- Emitting Devices. *Adv. Mater. Sci. Eng.* **2016**. https://www.hindawi.com/journals/amse/2016/4081697/.
10. Ma, Z. et al. Radio-Frequency Flexible and Stretchable Electronics (Key Note). In: *2016 China Semiconductor Technology International Conference (CSTIC)* **2016**; pp 1–4.
11. Seo, J.-H.; Ma, Z.; Zhou, W. Radio-Frequency Flexible Electronics: Transistors and Passives. In: *2014 IEEE Bipolar/BiCMOS Circuits and Technology Meeting (BCTM)*; **2014**; pp 107–114.
12. Semple, J.; Georgiadou, D. G.; Wyatt-Moon, G.; Gelinck, G.; Anthopoulos, T. D. Flexible diodes for radio frequency (RF) electronics: a materials perspective. *Semicond. Sci. Technol.* **2017,** *32* (12), 123002.

13. Petrone, N.; Meric, I.; Chari, T.; Shepard, K. L.; Hone, J. Graphene Field-Effect Transistors for Radio-Frequency Flexible Electronics. *IEEE J. Electron Devices Soc.* **2015,** *3* (1), 44–48.
14. Yonghui, Z.; Mei, Z.; Liang, H.; Du, X. Review of Flexible and Transparent Thin-Film Transistors Based on Zinc Oxide and Related Materials. *Chinese Phys. B.* **2017,** 26. DOI: 10.1088/1674-1056/26/4/047307.
15. Fan, F.-R.; Tian, Z.-Q.; Wang, Z. L. Flexible Triboelectric Generator. *Nano Energy* **2012,** *1* (2), 328–334.
16. Gieraltowska, S. et al. Properties and Characterization of ALD Grown Dielectric Oxides for MIS Structures. ArXiv11075401 Cond-Mat, Jul. **2011**.
17. Lin, H. K.; Hsu, W. C. Electrode Patterning of ITO Thin Films by High Epetition Rate Fiber Laser. *Appl. Surf. Sci.* **2014,** *308*, 58–62.
18. Sakamoto, K.; Kuwae, H.; Kobayashi, N.; Nobori, A.; Shoji, S.; Mizuno, J. Highly Bendable Transparent Electrode Using Mesh Patterned Indium Tin Oxide for Flexible Electronic Devices. In: *2017 IEEE 12th International Conference On Nano/Micro Engineered and Molecular Systems (NEMS)*; **2017**; pp 323–326.

CHAPTER 6

Flexible Substrates for Radio Frequency Electronics Inclined to 5G Technology

ARUN AGARWAL[1*], VARUN SHARMA[2], GOURAV MISRA[2], and SAURABH NARENDRA MEHTA[3]

[1]*Department of ECE, ITER, Siksha O Anusandhan Deemed to be University, Bhubaneswar, Odisha, India*

[2]*School of Electronic Engineering, Dublin City University, Dublin, Ireland*

[3]*Department of ECE, Vidyalankar Institute of Technology, Mumbai, Maharashtra, India*

ABSTRACT

Flexible electronics have been emerging since the development of nanoscale materials. These materials led to different ways of fabricating and manufacturing the product according to the radio frequency needs. It is estimated that the flexible electronics will boom and reach a level where ICs are in the current phase. The development of radio frequency started with radio frequency identification (RFID) communication with a very small range. Later on, these RFID devices were developed on the flexible substrates for applications such as security and identification. Rectifier diodes and complementary metal–oxide semiconductor (CMOS) circuits were the major components to be shifted to flexible substrates. The electronic circuits such as thin film transistor (TFT), capacitors, Schottky diodes, switches and diodes are studied as per application. After deciding the electronic circuits to build on, the next comes the substrate. The substrates such as plastics and e-textiles are the considered one due to

5G Green Communication Networks for Smart Cities. Devasis Pradhan, Rajeswari, Hla Myo Tun, Naw Khu Say Wah, & Thandar Oo (Eds.)

its flexible and electronic properties. The substrate should be able to withstand thermal properties, channel length, switching frequency of transistor. The active layer may be silicon membrane, small molecules and polymer, carbon nanomaterial, and metal oxide semiconductor with graphene and indium gallium zinc oxide (IGZO) being the most desired material. Fabrication techniques are considered based on substrate and active material such as screen printing, flexography, lithography and chemical etching and inkjet printing. Printing techniques have evolved due to availability of conductive, inductive and dielectric ink. Roll-to-Roll technique is widely used due to its low cost, large print area, and low-temperature process without the need of a clean room. Antenna design is also discussed with possibilities of using heterogeneous substrates, three-dimensional (3D) printing, and many more. The increasing research and developments are making the flexible electronic industry more versatile and ready for opaque and more robust flexible electronics.

6.1 INTRODUCTION

The ever-evolving electronics field is entering the stage where the need for miniature, low-power, and versatile devices is increasing. The demand is not only limited to consumer products but also in manufacturing industries, medical services, energy production, and many more. Like the way integrated circuit (IC) evolution led to many developments through all the industries, the next generation of electronics is the Internet of Everything (IoE). The demand for such can be fulfilled by integrating every other electronic sensor, device, or material with the internet. The initial requirement of making the device fast, lightweight, and most importantly small creates a big challenge for manufacturing industries to retain the device properties even with nanoscale dimensions. The applications of electronics in fields such as healthcare, light-emitting diode (LED), displays and many more demanded a need for implementing a large number of transistors in a constrained dimension. Right from displays where organic diodes were playing a huge role in IC growth and haptic feedback, different materials were researched to work with the exact needs of an application. Advanced manufacturing such as three-dimensional (3D), inkjet printing, and compatibility of substrates with flexible materials led to the emergence of flexible electronics. This field allowed a different electronic application to inherit its exact nature with the advantage of being flexible. Flexible applications in healthcare service such as bionic ears which make use of flexible electronics using thin film transistors along with pressure sensor array to replicate the original ear. Another healthcare application such as using textile fabrication for a sensor in bed sheets in the hospital to monitor the heat, pressure, and humidity of a patient.[1] The automobile industry also has many requirements for flexible electronics with the evolution of electric cars. The need for supercapacitors and thin-film battery technology.[1] Graphene in the field of touch and haptic feedback has allowed the device to be transparent as well as flexible. Graphene has also impacted the battery/electric storage industry by providing the fast-charging capability and longer life of battery cells.[1] Energy-harvesting industries also are making use of flexible electronics to acquire energy from the environment using nanoelectro-optomechanical systems. Thin film solar cells can harvest twice the energy compared to the one by bulk solar cells. Mobile power harvesting systems are growing with the increase of smart grid technology and electrical heavy appliances.[1] However, for all these devices there is a basic need for communication for interconnectivity between electronic appliances.

Communication can be achieved through wireless systems which can also be fabricated to be efficient and flexible. The basics of a communication channel consist of a transistor and a receiver. These antennas are fabricated on a flexible substrate to work in the range of many frequencies. The radio frequency applications range from Radio frequency identification (RFID) to near-field communication (NFC) and cell phone antennas. Basic communication involves the transmitter sending data through an antenna which is then received by the receiver with the postprocessing process. RFID applications work on this principle which works using a unique identification (ID) generated by one of the sources.

There can be two types of devices known as inactive or passive devices. In the passive device, the tag within which the code is saved is powered by the help of the reader's electromagnetic waves. These waves are rectified and then the power is passed on to the complementary metal–oxide semiconductor (CMOS) circuit to get the encoded code. The code is later on received by the reader. For active devices, the tag is powered by its independent battery source and sends the code back to the receiver. These devices generally run at a frequency of 13.56 MHz with a distance range of

10 cm.[2] Similar applications can be found in mobile phones which utilize NFC providing a bi-directional data flow. The range of these devices is one of the factors that determine the performance of the tags. RFID is used many applications such as security, identification tag, shopping tags, access control to communicate, or pass security keys.[2]

In this paper different types of substrates, circuits, fabrication techniques, and antenna designs. The aim to achieve flexible electronics with low-cost production, efficiency, good strain properties, and frequency parameters are discussed in the following.

6.2 ELECTRONIC CIRCUITS

Different applications that work in a range of chinless to ultrawideband frequencies have specific circuit requirements. The communication medium between devices also varies such as electromagnetic and magnetic, or electric.[3] Read range defines the parameters related to medium, antenna size, and power management. The rise of applications in the HF range and UHF range is leading to improvement in the performance of rectifiers of the circuit. The rectifier impacts the frequency performance of the device and manufacturing a flexible rectifier along with the diodes involved distorts the expected frequency.[2] Rectifiers, logic circuitry, load modulator, sequence generator, and CMOS are the components to look into while considering a radio frequency circuit and further on to make it flexible for high-frequency applications.

6.2.1 RECTIFIER

A diode along with a capacitor is used in a circuit to achieve a direct current (DC) wave using an alternating current (AC) input. The current flow is permitted in one direction by the diode to conduct half wave signal which is then sent to the capacitor to charge. Such a simple circuit has small ripples which are related to capacitor RC time.[3] To refine the output wave, several diodes along with capacitors can be arranged for a voltage multiplier circuit. Again, with an increase in a number of components, the deformable capability can be a hit or miss in certain conditions of strain.

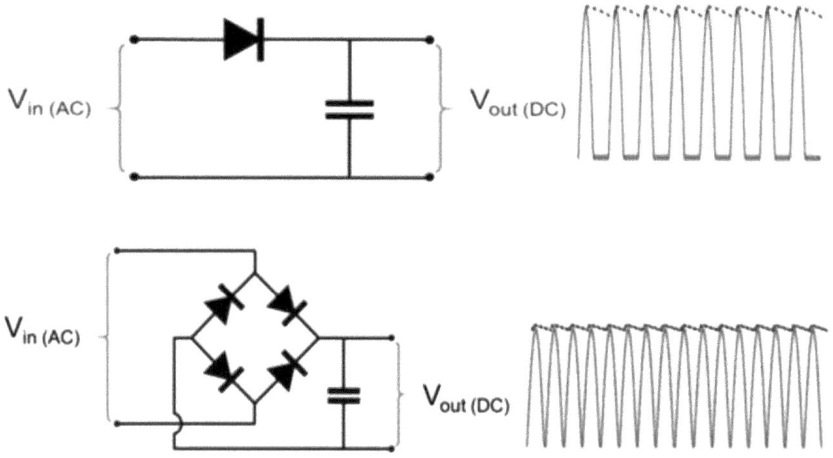

FIGURE 6.1 Diode outputs.

Source: Reprinted from Ref. [2]. https://creativecommons.org/licenses/by/3.0/

6.2.2 SCHOTTKY DIODE

Considering a low-voltage application, the Schottky diode is preferred due to its high current irrespective of bias direction and lower switching voltage.[2] Being a metal semiconductor diode, there is an observed bending near the semiconductor junction due to a difference in the Fermi level. The charge carrier at this junction has a reduced barrier height to pass through, hence the lower switch on compared to other diodes.[4] Comparing the switching performance to that of *p-n* junction diodes wherein interfaces stores the charge before discharging and hence the delay.[2]

6.2.3 TRANSDIODE TRANSISTOR

The transistor is used as a replacement for the diode in the conversion circuit. There are two ways where a transistor can be used for rectifying, one by providing the variable gate voltage and the other in a diode configuration with a connection between the gate and drain.[2] Comparing the radio frequency performance it was found that Schottky diodes outperformed transdiode transistors by a good value. Despite having a performance drop, transistors are still preferred as fabrication is more flexible and in line with other logic circuitry and rectifiers.[2]

For successful RF communication, the sender and receiver should be compatible with a particular frequency along with the power management requirements. To further classify the features of RF communication there are other factors such as antenna type, operation type, data quantity (authentication, cryptographic, and reprogrammable), frequency range, and data transfer transponder.[3] Considering these requirements of electronic circuits and antenna configuration upcoming devices should be flexible. To make it flexible, substrate and material deformable capabilities are researched further.

6.2.4 THIN FILM TRANSISTORS

Manufacturing thin film transistors on a flexible substrate requires very high carrier mobility along with fabrication techniques that can be compatible with flexible substrates.[5] These thin film transistors (TFTs) are known as active devices and a suitable Silicon Nan membrane is considered for a wide number of applications. The fabrication of this TFT is first constructed using a solid substrate and later on, after etching the layer is transferred to a flexible substrate.[5]

6.2.5 CAPACITORS AND INDUCTORS

To form flexible capacitors, the typical process of using plasma-enhanced chemical vapor deposition was not compatible with the plastic substrates due to high temperature. Using the e-beam evaporation system 200-nm silicon monoxide layer is deposited in a patterned way. Over this dielectric layer, an evaporation of metal layer is used which comprises of material such as Au and forms the top electrode of the capacitor.[5] For inductors, the inter-metal dielectric constant is low leading to low values of parasitic capacitance. This low ratio of dielectric results in higher frequencies for inductors.[5]

6.2.6 SWITCHES AND DIODES

The fabrication process of diodes and radio frequency switches is very similar to that of TFT. This makes it easy to fabricate diodes along with

TFT.[5] Ion implantation of *p*-type doping is an extra required step for switches. The materials used in the process are Si NM and Ge NM, both being worthy components for many manufacturers. The turn-on voltage observed with Si was around 0.75 V and for germanium, it was around 0.45 V. The frequency response of Ge NM was much better compared to Si NM diodes. Another plus point of using Ge was the minimum change in frequency parameters whenever strain was applied making it more suitable for flexible electronics.[5] Despite having so many advantages, Si was the most preferred material as it was very much compatible with active components such as CMOS while inhibiting near-to-Ge electronic and mechanical properties.[5]

6.3 FLEXIBLE SUBSTRATE

The evolution of flexible electronics starts with the very basic substrates utilized in production. Flexible substrates give a new dimension to all the applications ranging from micro- and macroelectronics. Applications in the fields of heavy industries, medical, manufacturing, wearables, display, fabrication, and many more will be evident with new ways of making substrate flexible, lightweight, high speed, and inexpensive.[6] The major challenge of placing circuits on a flexible surface, developing sensors that can be twisted or electronic papers can be achieved by making the substrate flexible.[7] Substrates which are perfectly suited for the flexible applications should possess properties such as thermophysical nature, resistance to solvent, ability to develop electronic grade materials at a lower temperature. Also, postprocessing techniques such as handling and packaging.[7]

For a substrate to be categorized as flexible it should withstand extreme bending conditions and not affect the components within or the performance over time.[8] The material to be used for substrate was important for its properties inhibited and a wide topic of research. Substrate material such as stainless steel has a high density of mass which makes it heavy for portable device, also due to its very poor flexible nature, it becomes difficult to use this for flexible applications.[1] There is research in context with fluidic substrate layer. Using eutectic gallium indium alloy which is a liquid metal improves the deformity strength of the material making it a more flexible and lasting shell life. EGaIn has properties such as self-healing which helps to extend the durability of the material after maximum deformation, and twists also its nature of being liquid at room temperature helps.[9] Another flexible substrate, the fabric can also be utilized in the fabrication of devices. Electric textile can be used to fabricate antennas that are flexible due to benefits such as its high resistance, and low resolution of less than 1 mm compared to metallic materials.[10] There are ways in which threads are sculpted in different ways leading to the embroidery of different electrical components on the surface of textiles such as metal filaments bundle and many more.[10]

Paper a flexible, organic material that can also be used as a substrate for fabrication. The benefits related to it are hugely due to its easy/low-cost production techniques such as inkjet printing and direct wire methods. Properties related to paper as a substrate such as a nanoparticle ink (polyvinylpyrrolidone, functionalized silver particle, diethylene glycol).[11] Plastic is a preferred substrate due to its lightweight, variable fabrication temperature for different types of plastics and deformable. Glass transition temperature is the factor to be considered while selecting a particular plastic substrate as after a particular temperature, materials get deformed making it difficult for photolithography.[1] Polyethylene terephthalate (PET) is widely used in various optoelectronic applications and is a worthy replacement for conventional glass, PET with film composed of crystallized indium tin oxide improves the fabrication process.[12,13] Certain features can be enhanced as per the application requirement wherein the substrate required should be opaque and more reflective. Under these scenarios, PET is aluminized with the help of metal evaporation over it.[13] Device performance depends on factors such as channel length, switching frequency of the transistor, and mobility related to the material.[14] With these properties, if the material is recyclable it is more sustainable.

To reduce the size of the substrate and make it more efficient, research was conducted to utilize the existing flexible substrates and integrate them with certain electrode properties that support the upper layer circuit. Heterojunction Silicon with flexible substrates demands an annealing temperature of more than 150°C, thermal stability at such temperature is also a requirement for roll-to-roll fabrication techniques.[15] One such example of a hybrid substrate is the FEAM-flexible PET substrate Ag mesh embedded within the material. It is widely used in perovskite solar cells.[15] Polyimide substrates are also considered in thermal/photonic annealing as they can easily withstand a temperature of 400°C.[16] Substrates were developed with waveguide components integrated within them.[6] This technique was compatible with plastic and paper substrates hence also achieving its deformation properties. Paper-based substrate integrated waveguides were able to achieve frequency in the range of 3–10 GHz when used as a filter in a particular setup.[17] Utilizing this concept of integrated waveguides with plastic, a frequency range of 5 GHz was achieved though there was a drop after bending stress.[6] With the help of such substrate-integrated waveguide, component costs can be kept in control along with a scope to develop more features related to it. Considering all the above-discussed features, a substrate can be picked up based on dielectric constant, frequency range and metallic line conductivity.[8]

6.4 MATERIAL FOR ACTIVE LAYER

After selecting the electronic components required for the design and then determining the flexible substrate to make it deformable, next is the material that will be used, for example, semiconductor resting on top of the substrate. The material should not only be compatible with the substrate used beneath it but also support the manufacturing process required to adopt it. Factors such as fabrication and sustainability are also taken into consideration before picking a material. The annealing temperature for different substrates varies such as for PET less than 150°, polyimide requires below 300°C hence keeping it low to be considered for the particular material.[2] Classifying different materials based on flexibility, mobility, printability, maturity, and frequency helps to narrow down to a particular requirement. The graph shown in Figure 6.2 states that carbon nonmaterials, flexible silicon, and Molybdenum disulfide are the material with a frequency of more than 109 Hz and mobility ranging from 101 to 104 $cm^2/V/s$.

6.4.1 SILICON NAN MEMBRANE

A sheet of semiconductors which are in nanoscale dimensions and derived from silicon-on-insulator is a popular choice used for nanostructures. Sheets in the range of 10–100 nm can be obtained by only etching the buried oxide part of SOI leading to deformable crystals.[18] These materials are known as silicon nanomembranes. Characteristics such as strain,

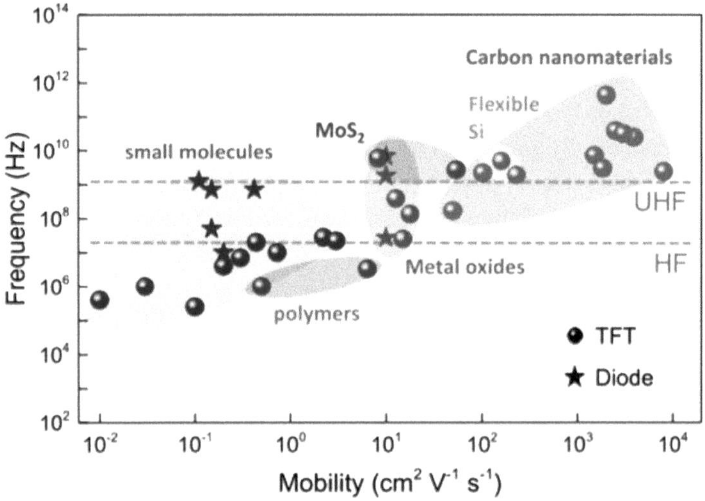

FIGURE 6.2 Comparison of materials with respect to mobility.

Source: Reprinted from Ref. [2]. https://creativecommons.org/licenses/by/3.0/

stacking, and deformation related to Si NM make it a research topic in waveguides and photonic applications. Changes in band structure due to strain results in high mobility and flexibility make it adaptable to any shape.[18] The maximum oscillation frequency of 7.8 GHz and a cut-off frequency of 2.04 GHz were obtained by utilizing the silicon nanomembrane on a plastic substrate.[19] Frequency response can be varied by changing the distance between source to the gate and drain to gate.[19] An alternative to silicon, germanium was also considered due to its low turn-on voltage and higher switching speed. Germanium nanomembranes fabricated on plastic substrates yielded insertion loss around 1.3 dB for 30 GHz making it a good choice for RF/microwave applications.[2,20] Cellulose nanofibrillated fiber (CNF) substrate along with GaAs Schottky diodes and capacitor fabrication using nanomembranes have applications related to a stable Wi-Fi router buildup.[2] Techniques such as nanoimprinting lithography allowing 100 nm channel length to fabricate fine patterns. A trench channel etching approach to build by TFT using silicon nanomembranes achieved an oscillation frequency of 38 GHz. To reach this frequency of RF transistor, NIL was utilized and a gate length of 2 was built.[21]

Another form of silicon approach where utilizing silicon microparticles was brought into consideration. Screen printable semiconductor composite made up of silicon microparticles and form a diode by placing it between electrodes resulting in diodes.[22] The frequency response of this diode achieved was 1.6 GHz.[22]

6.4.2 SMALL MOLECULES AND POLYMERS

The benefits of organic semiconductor materials such as being stretchable, deformable, and lightweight give them an upper hand compared to existing semiconductor materials.[2] Also, fabrication advantages such as processing at low temperature along with compatibility with printing techniques make it desirable for flexible electronics applications.[2] The introduction of small molecules in the active semiconductor layer has gained importance as it enhances the carrier field mobility. Organic material such as pentacene which is similar to polyacene is a recognized constituent used for the fabrication of organic TFTs.[23] In the procedure in which the pentacene is used such as using a stream of inert gas to harvest pentacene films, the carrier mobility observed is around 3 cm^2/Vs.[23] Fabrication of the Schottky diode with pentacene led to great frequency improvements. Schottky diodes were built in a 3-layer structure with one forming a Schottky barrier using Al, PEDOT-PSS and the other forming an ohmic contact using Au, PEDOT–PSS with pentacene.[23]

Organic macromolecules such as conjugated polymers known for their varying single–double bond design have applications in flexible electronics providing more flexibility compared to small molecules.[2] Also being a polymer it is soluble and makes it a good candidate for solvent-based fabrications such as solution process, inkjet etc.[24] Comparing it to small molecules it has low mobility and rectifiers based on polymers were able to achieve 13.56 MHz by varying the thickness of the diode.[25] Polymers such as 3-hexylthiophene (P3HT) and 3,3-didodecylquaterthiophene (PQT-12) were used for the fabrication of diodes generally for rectification circuits with an output of 4 V DC for ±10 V AC.[26]

6.4.3 CARBON NANOMATERIALS

Different materials such as crystalline diamond, diamond-like carbon, graphene, and carbon nanotubes possess exceptional conductance and high mobility properties.[27] These materials are also gaining importance in several electrochemical, optoelectronic, and plasmonic applications due to their mechanical, thermal, and electronic properties.[2]

Carbon nanotubes CNT is one of the materials that have high breakdown voltage due to its thermal stability and low quantum capacitance with respect to its dimensions.[2] Dimensions, structure, and diameter decide the electronic properties of CNTs and the arrangement of nanomaterial in terms with density impacts the performance of the device.[28] Single-walled nanotubes are classified into two types—semiconducting and metallic nanotubes. TFTs and field effect transistors (FETs) are fabricated with a mixture of both semiconducting and metallic types utilizing their intrinsic properties and charge mobility.[28] CNTs can achieve mobilities of 105 cm^2 V^{-1} s^{-1} and up to 80 GHz cut-off frequency related to intrinsic current gain.[2] Fabrication of these materials takes place via chemical vapor deposition process wherein metal crystals are placed on top of the substrate at a temperature of 900°C.[28] During this CVD process, the alignment of nanotubes can be controlled by altering the gas intake, electric field, and material interaction with the substrate.[28]

Another method, methane CVD is also utilized to obtain a controlled combination of semiconducting and metallic tubes for specific requirements. Both these types have certain disadvantages related to it such as semiconductor wear off due to its rapid switching requirement and the metallic part impacts the gain and frequency of the material.[28]

The manufacture of CNT using CVD and other processes is the most common method of fabrication with the help of rigid substrates. The challenge is to manufacture CNT on top of flexible substrates that don't allow processing at temperatures above 150–300°C. Polyimide substrates were used to fabricate CNT by dispersing it in the solution and bringing it together through solvent.[2] Spin-based or inkjet-based fabrication of CNT over polyimide achieved operation frequency of 5 GHz.[2]

Similar to CNT, another carbon nanomaterial is being researched due to its exceptional electronic and mechanical properties known as graphene. Possessing abilities similar to that of a CNT, there is no quantization of conductance leading to very high intrinsic cut-off frequencies of around 427 GHz.[2] It was also observed that the carrier mobility of graphene was far beyond around 200000 cm^2 V^{-1} s^{-1} using a single-layer graphene.[29] Graphene used for diode applications have also some improved results such as current restrictions to mA and considering the impedance factor 65 GHz frequency range was achievable.[2] Initially, graphene was extracted by shedding layers through a mechanical process but with the latest development graphene was manufactured in a process such as CVD and later on inkjet, roll methods. The development of graphene over PEN and PET substrates for TFT manufacture led to a frequency range of 2.5–13 GHz. Utilizing polyimide substrate for graphene TFTs resulted in a major impact on carrier mobility with additional properties to resist water and high strain values.[2] It was also observed that charge carrier scattering and the mobility of carriers can be impacted by the underlying substrate material.[29]

Alternate methods of graphene developments such as using graphene ink, extracting of graphene directly into solution resulting in graphene oxide solvent and later exfoliating it to obtain the desired structure over substrate were observed. The absence of a band gap also makes it incompatible with many applications especially rectifying diodes.[2] Graphene oxide's gets easily distributed in water which makes the fabrication process to cover large area.[30] The insulating nature of graphene can be obtained when it is fully oxidized resulting in a band gap of 2.4 eV. Using optical photolithography and making use of oxygen to control the semiconductor properties of graphene helps in the fabrication of diodes.[30] With annealing temperatures around 100° C and graphene oxide at the expense of lower charge mobility, the rectifier was able to generate a 2-mV signal for a 26 GHz frequency.[2] Its liquid nature in the initial steps of fabrication made it more compatible with low-cost, high-scale printing techniques. Carbon nanomaterials are excellent candidates for optimal frequency, mobility, and flexibility-related applications.

6.4.4 METAL OXIDE SEMICONDUCTOR

Wide band gap semiconductors are used for many applications where an electrical conductivity change was required. Materials such as SnO_2, TiO_2, In_2O_3, and ZnO were used in great interest for optoelectronics, TFTs, and later flexible electronics. These materials lack uniformity due to which research was shifted to indium gallium zinc oxide (IGZO).[2] IGZO retained the properties of metal oxides while being uniform and providing excellent mobility carriers. Also, its wide band gap properties make the material optically transparent and conducting/insulating properties.[2] It was also used in RFID applications due to its low voltage switching capability. Logic gate configuration in RFID which is working based on IGZO IC requires a mere 20 μW which was feasible from a distance of 70 mm and receiving an AC signal of 40 mW.[31]

The performance of IGZO can also be controlled by varying the layer thickness along with large deposition methods such as sputtering wherein the oxygen ratio is varied.[4] Schottky diode manufactured using IGZO can range up to a frequency of 6 GHz, a performance that is obtained by carefully altering the fabrication process.[4] To adjust the diode barrier height and ideality factor postannealing methods were implemented along with the use of fast laser annealing which did not affect the substrate by keeping a stable temperature well under control.

With such tweaks, 0.6 V output of rectifier was observed for ±2 V AC signal.[4] In the diagram shown in Figure 6.3, graphene and IGZO are compared based on their properties and are widely used materials for flexible electronics.

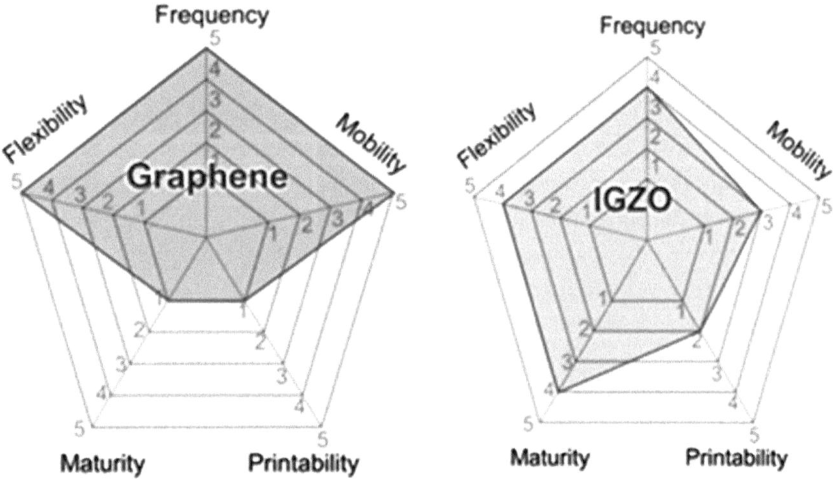

FIGURE 6.3 Properties of graphene and IGZO.

Source: Reprinted from Ref. [2]. https://creativecommons.org/licenses/by/3.0/

6.5 MATERIAL FOR ACTIVE LAYER

The technique of manufacturing flexible electronics impacts a lot of factors such as cost, device properties, device morphology, and quantity. Considering these factors and selecting the above-mentioned active layer material and substrates few methods are commonly used for fabrication.

For printing methods, there are types of ink such as dielectric, semiconductive, organic, and piezoelectric conductive inks which help to adjust the different properties of active layer material as per the need.[32] A few methods are described in the following.

6.5.1 SCREEN PRINTING

Electronic manufacturers prefer a low-cost simple method for fabrication, screen printing being one of them. The process involves a woven screen which acts as a printing head made up of varying densities and thickness. This screen is then placed on top of the substrate. To inject the ink on the top of a substrate, the squeegee blade is used to produce pressure on the screen which then leaves the ink onto the substrate through the different thicknesses of pores.[8] The exposed area and the pores lead to the desired print requirements in the fabrication using materials such as polyester. The process involves a conductive ink which is applied through an already masked screen and thermally cured on the substrate.[8]

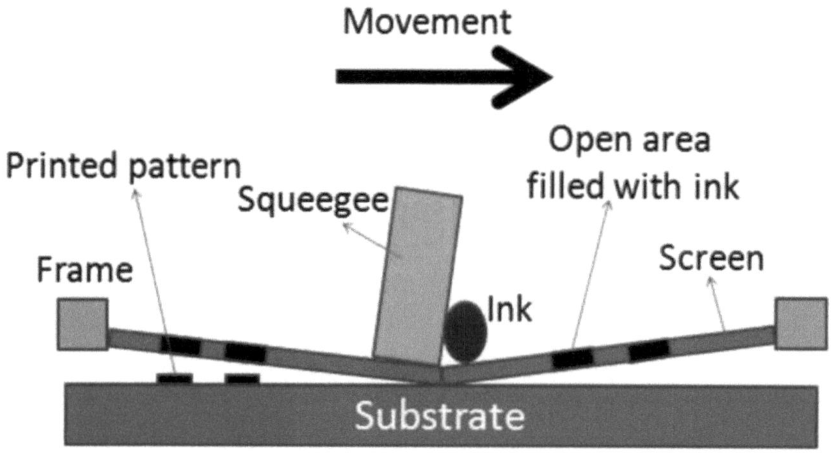

FIGURE 6.4 Screen printing process.

Source: Reprinted with permission from Ref.[8]. Copyright © 2013 The Author(s). Licensee IntechOpen. https://creativecommons.org/licenses/by/3.0/deed.en

The printing process is considered to be an additive process where the print is masked on the substrate. Other methods similar to this process are also developed and researched such as cylinder, rotary, and flatbed where the process of printing conductive ink remains the same. For cylinder screen printing, the substrate is rotating and the pattern is then transferred as screen rolls.[8] In flatbed assembly the screen is flat and in rotary assembly, the same screen is rolled. Inside the cylinder, squeegee and ink are present to deposit in the pores and markings. Impression cylinders are used to create adequate pressure on the screen onto the substrate.[32] Factors which are still needed to be considered are the resolution of print, layer thickness, and passes. Ink viscosity can be affected during the process due to thermal processing requirements.[8]

6.5.2 LITHOGRAPHY AND CHEMICAL ETCHING

One of the methods used to develop PCB is now widely used to fabricate patterns known as chemical etching. Organic polymers such as photoresist have a particular reaction in the presence of ultraviolet (UV) rays. If the layer which is exposed gets soluble then the type of resist is considered positive and if solubility is very low then negative.[8] A positive photoresist is the first choice in this type of fabrication as with negative there is swelling observed. With the help of photoresist for metallic patterns can be obtained using photolithography and a particular area can be etched using etchants. This process of chemical etching can give a high-precision pattern.[8] The technique used is considered to be a complicated, clean room process with many postprocessing steps involved increasing the overall cost.

6.5.3 FLEXOGRAPHY

Similar to screen printing, a commonly used printing method is flexography. The working principle of this process is to get the ink patterned the substrate through a cylinder. Patterns to be printed are created using the printmaking process which is used as pattern plates built using rubber.[32] These pattern plates get attached to the plate cylinder. Another roller present next to this cylinder is an anilox roll. The role of an anilox roll is to provide ink through its engraved cups according to print size and ink.[32] Antenna manufacturers consider this technique due to its large throughput, reduced cost, and fine resolution with requirements such as low-viscosity ink.[8] Printing pressure, duration of a print, quality of substrate surface, anilox roll variant, printing plate, and hydrophobicity of substrate are the constraints that can enhance/degrade the print quality.[32]

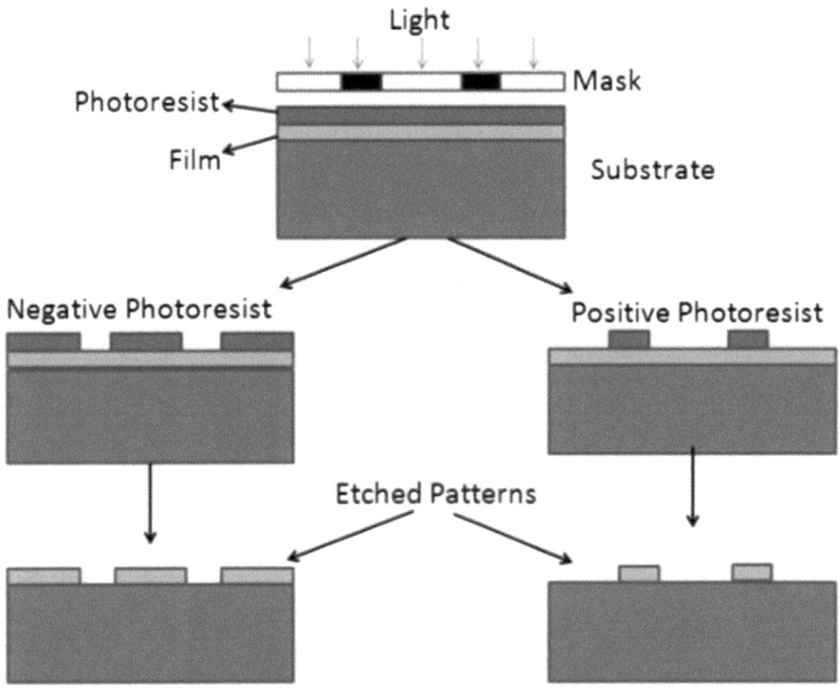

FIGURE 6.5 Chemical etching process.

Source: Reprinted with permission from Ref.[8]. Copyright © 2013 The Author(s). Licensee IntechOpen. https://creativecommons.org/licenses/by/3.0/deed.en

Gravure printing works, similarly, with just two cylinders. The plate cylinder and anilox roll are replaced by one gravure cylinder. The print plate is now taken care of by gravure and the pattern is directly carved on it. Different methods are available to engrave the pattern on gravure such as electrochemical, laser, and chemical etching. Ink quality along with gravure cylinder properties needs to be very finely detailed as it impacts the print quality.[32] Uniform structures of print are achieved by maintaining the ink viscosity which being improper can lead to ink spilling from the cylinder. Roll printing is being used in production line assemblies and resolutions up to 1270 dpi prints are very much feasible.

6.5.4 INKJET PRINTING

As with the latest developments in circuit printing and fabrication, the types of ink available are also evolving. Conductive inks can be used as droplets on top of a substrate to form a pattern. This process of ink deposition via inkjet material printers is known as inkjet printing.[8] Nano-structural materials are used to generate conductive inks such as gold, copper, silver, nickel nanoparticles, and CNT.[6] Increasing nanoflakes content can lead to an increase in the clogging rate along with the viscosity of the ink. To get rid of the solvent from conductive material after printing by curing the substrate. Sintering for a long duration of time via high-power UV helps to achieve good conductivity of the print.[6] Insulating inks are also used in this printing solution for the dielectric layer. These techniques are divided into two groups known as continuous technique and drop-on-demand technique. The continuous inkjet printing inkjet technique involves a high stream of ink passed through a nozzle and then with the help of electrostatic plates the flow is directed to form a particular pattern.[6]

The drop-on-demand technique follows a similar process where droplets controlled by the electric field are used to print precise patterns. Piezoelectric or thermoelements are used to create a pressurized pulse which in turn drives a drop through the nozzle. The working of a thermal inkjet printer involves a vapor bubble which is generated with the help of heated plates at temperatures of around 300°C.[6] Factors such as particle size, surface tension, print head properties, substrate type, and topology along with the viscosity of the liquid can define a high print quality. The printing process is monitored and coordinated using a computer that too without the need for a clean room.[8] Another type of jet printing known as aerosol printing based on mesoscale material printing without the need for a mask. With the help of heat, ink is broken up into very tiny droplets with diameters ranging in micrometers. These droplets are then made to pass through a gas stream to a print head. The gas stream is compressed to 10 μm while passing through the exit nozzle. A print resolution of 5 μm can be achieved through aerosol printing.[6] Inkjet printing techniques is a widely used method by many manufacturers as it provides the facility to control the ink droplet printing for a specified design and without the need of a clean room environment in less duration.[8]

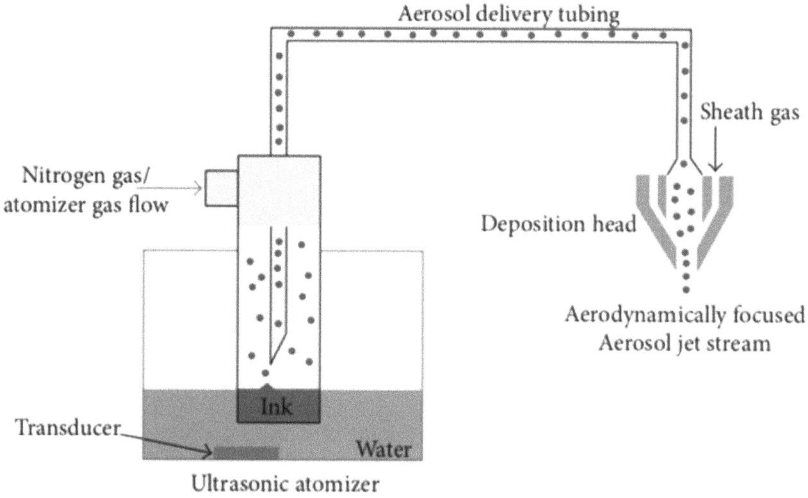

FIGURE 6.6 Aerosol jet printing.

Source: Reprinted with permission from Ref. [6]. https://creativecommons.org/licenses/by/4.0/

6.6 ANTENNA TYPES AND DESIGN

The final approach to getting flexible electronics for radio frequency applications is the antenna design and methods implemented to manufacture one. The final design and properties depend on the electronic circuit to be used, the flexible substrate good enough for the strain characteristics, the active material on top of the substrate, and finally the fabrication technique to construct the design. A few antenna designs implemented in different ways are mentioned in the following.

6.6.1 *MICROSTRIP PATCH ANTENNA*

The antenna developed for any application can be evaluated based on 2D radiation pattern, impedance matching, and right-hand circularly polarized parameters.[33] Wideband antenna is being considered with a 4 × 4 patch array to achieve high gain. To achieve this substrate material considered is PET, the inkjet process for printing the desired design, and finally photonic curing. To make the antenna flexible and to make it compatible with roll-on printing, PET was considered to be a good choice owing to its polyester nature.[33] Being a 4 × 4 array construction choice, the array design was passed on to the printer. With the help of inkjet printing techniques, the design was printed onto the substrate uniformly using conductive inks. The next step in this process is to cure, to get the ink settled down to the specified depth, conductivity, and if required before going for the next layer of the print. Curing types such as photonic were used wherein high temperatures using pulsed lights.[33]

6.6.2 HETEROGENEOUS SUBSTRATE

The types of antenna design wherein the substrate is modified with some other material to enhance the radio frequency property. Substrate such as plastic, paper, and textiles are very commonly used for substrate-integrated waveguides. Utilizing paper-based integrated substrates operation frequency of 10 GHz was achieved.[6] Plastic-based substrate materials were considered in applications of energy-harvesting systems. Textile-based integrated waveguides involve a process wherein the conductive polymers are wrapped on the top of the textile substrate. Holes were cut upon the textile for the design. It always performed even with enough strain and was worthy enough to be used along with clothes as a wearable device.[6] Another example of a heterogeneous design is where the substrate is 3D printed to achieve the desired bandwidth and gain. To maintain optimal mechanical properties and radio frequency parameters, the substrate can be modified while it is being manufactured. Substrates can be constructed using 3D printer technology and thereby modifying the substrate material with the printing layers. The author in Ref. [34] demonstrated an antenna, wherein NinjaFlex substrate was considered but due to low radio frequency performance at a particular dissipation factor angle. By using the latest 3D printing technology, the proportion of NinjaFlex and acrylonitrile butadiene styrene (ABS) was constructed layer by layer leading to optimal mechanical and radio frequency parameters.[34] The proportions of ABS can alter the radio frequency components and a desired heterogeneous substrate was obtained. It was also observed that with the rise in ABS proportion, bandwidth was impacted but increase in gain was observed.[34]

6.6.3 MEMBRANE PHASED ANTENNA

Phased array antennae used for applications such as space, and radar industries are comparatively heavy and to solve this problem active membrane phased array is considered.[6] Construction of this membrane is as such with two layers, one comprising of radiating patches and the ground plane of these patches on the other. Author in Ref. [35], used a 50-μm Pyralux polyimide flexible membrane substrate for the fabrication. A separate hybrid module holds the Transmit/Receive module and is connected to the membrane array.[35] The Traditional method of etching is used for the interconnects and designs to be implemented.[6]

6.6.4 RADIO FREQUENCY ALTERING ANTENNA

It was observed that the radio frequency properties of the antenna can vary due to deformation or bend and there is another antenna that does not change its radio frequency properties even after mechanical changes. To observe this property an accordion-based monopole with foldable parameters was used to operate at 2.4 GHz. The frequency response of the material was observed for every deformation from 10° to 60° and it was found that a frequency of up to 3.47 GHz was observed.[36] Similarly, an antenna based on an e-textile design such as a cotton substrate was considered. This antenna was also made to perform at 2.4 GHz and then the textile was deformed at several angles. Comparing this to the previous one there was no noticeable frequency change (gain/reduction) for every angle.[36] Hence, it was possible with the help of certain substrates to achieve a fixed frequency and a variable frequency for others at different deformation angles.

6.7 CONCLUSION

The demand for flexible electronics was taking an exponential curve and with that demand, the recent developments in each part of the material right from the substrate to the antenna was also increasing. Different types of materials that can be used to build up a radio-frequency application such as rectifiers, diodes, TFT, switches, and transdiode transistors were considered. The flexible substrates available right from plastic, paper, textile, and liquid texture were considered. However, PET still is the most preferable choice. The latest trend of the 3D printer, allowed the manufacture to construct heterogeneous materials. Active layer components were discussed showing different mobility charge speed and strain capacity. Graphene and IGZO were the widely used due to its exceptional electronic and radio frequency properties. Finally, the antenna design was decided to make a microstrip or array as per the need. Different aspects were considered to build a proper radiofrequency application with the help of a flexible device.

KEYWORDS

- **flexible electronics**
- **radio frequency applications**
- **flexible substrate**
- **inkjet printing**
- **waveguide antenna array**
- **e-textile substrates**
- **active layer materials**
- **graphene**
- **heterogeneous substrate**

REFERENCES

1. Nathan, A. et al. Flexible Electronics: The Next Ubiquitous Platform. *Proc. IEEE* **2012,** *100*. Special Centennial Issue, 1486–1517. DOI: 10.1109/JPROC.2012.2190168.
2. Semple, J.; Georgiadou, D. G.; Wyatt-Moon, G.; Gelinck, G.; Anthopoulos, T. D. Flexible Diodes for Radio Frequency (RF) Electronics: A Materials Perspective. *Semicond. Sci. Technol.* **2017,** *32* (12), 123002. DOI: 10.1088/1361-6641/aa89ce.
3. Finkenzeller, K. *RFID Handbook: Fundamentals and Applications in Contactless Smart Cards, Radio Frequency Identification and Near-Field Communication*, 3rd edn.; p 480.
4. (PDF) Flexible Indium–Gallium–Zinc–Oxide Schottky Diode Operating Beyond 2.45 GHz. ResearchGate. https: //www.researchgate.net/publication/280030346_Flexible_indium-gallium-zinc-oxide_Schottky_diode_operating_beyond_245_GHz (accessedon 28 Nov 2019).
5. Seo, J.-H.; Ma, Z.; Zhou, W. Radio-Frequency Flexible Electronics: Transistors and Passives. In: *2014 IEEE Bipolar/BiCMOS Circuits and Technology Meeting (BCTM)*; **2014**; pp 107–114. DOI: 10.1109/BCTM.2014.6981296.
6. Monne, M. A.; Lan, X.; Chen, M. Y. Material Selection and Fabrication Processes for Flexible Conformal Antennas. *Int. J. Antennas Propagation* **2018**. https://www.hindawi.com/journals/ijap/2018/9815631/ (accessed on 27 Mar 2020).
7. Nathan, A.; Chalamala, B. R. Special Issue on Flexible Electronics Technology, Part 1: Systems and Applications. *Proc. IEEE* **2005,** *93* (7), 1235–1238. DOI: 10.1109/JPROC.2005.851525.

8. H. R.; H. M.; A. I. Design, Fabrication, and Testing of Flexible Antennas. In: *Advancement in Microstrip Antennas with Recent Applications*; Kishk, A., Ed.; InTech, **2013**. http://www.intechopen.com/books/advancement-in-microstrip-antennas-with-recent-applications/design-fabrication-and-testing-of-flexible-antennas (accessed on 29 Mar 2020).
9. Kubo, M. et al. Stretchable Microfluidic Radiofrequency Antennas. *Adv. Mater.* **2010,** *22* (25), 2749–2752. DOI: 10.1002/adma.200904201.
10. (6) (PDF) Textile Based Embroidery-Friendly RFID Antenna Design Techniques. ResearchGate. https://www.researchgate.net/publication/333345520_Textile_Based_Embroidery-Friendly_RFID_Antenna_Design_Techniques (accessed on 29 Mar 2020).
11. Anagnostou, D. E.; Gheethan, A. A.; Amert, A. K.; Whites, K. W. A Direct-Write Printed Antenna on Paper-Based Organic Substrate for Flexible Displays and WLAN Applications. *J. Disp. Technol.* **2010,** *6* (11), 558–564. DOI: 10.1109/JDT.2010.2045474.
12. Lee, S. J., et al. Flexible Indium–Tin Oxide Crystal on Plastic Substrates Supported by Graphene Monolayer. *Sci. Rep.* **2017,** *7* (1), 1–8. DOI: 10.1038/s41598-017-02265-3.
13. (6) (PDF) PET as a Plastic Substrate for the Flexible Optoelectronic Applications. ResearchGate. https://www.researchgate.net/publication/224880095_PET_as_a_plastic_substrate_for_the_flexible_optoelectronic_applications (accessed on 30 Mar 2020).
14. (6) (PDF) Ultra-Thin Chips for High-Performance Flexible Electronics. ResearchGate. https://www.researchgate.net/publication/323758164_Ultra-thin_chips_for_high-performance_flexible_electronics (accessed on 27 Mar 2020).
15. Li, Y. et al. High-Efficiency Robust Perovskite Solar Cells on Ultrathin Flexible Substrates. *Nat. Commun.* **2016,** *7* (1), 1–10. DOI: 10.1038/ncomms10214.
16. Arapov, K. et al. Graphene Screen-Printed Radio-Frequency Identification Devices on Flexible Substrates. *Phys. Status Solidi RRL—Rapid Res. Lett.* **2016,** *10* (11), 812–818. DOI: 10.1002/pssr.201600330.
17. Moro, R.; Bozzi, M.; Kim, S.; Tentzeris, M. Novel Inkjet-Printed Substrate Integrated Waveguide (SIW) Structures on Low-Cost Materials for Wearable Applications. In: *2012 42nd European Microwave Conference*; **2012**; pp 72–75. DOI: 10.23919/EuMC.2012.6459165.
18. Lagally, M. G. Silicon Nanomembranes: Opportunities for New Si Functionalities via Strain, Flexibility, and Layering. In *LEOS 2008 - 21st Annual Meeting of the IEEE Lasers and Electro-Optics Society*; **2008**; pp 632–633. DOI: 10.1109/LEOS.2008.4688777.
19. (10) (PDF) 7.8GHz Flexible Thin-Film Transistors on a Low-Temperature Plastic Substrate. ResearchGate. https://www.researchgate.net/publication/253597256_78GHz_flexible_thin-film_transistors_on_a_low-temperature_plastic_substrate (accessed on 09 Apr 2020).
20. Qin, G. et al. Fabrication and Characterization of Flexible Microwave Single-Crystal Germanium Nanomembrane Diodes on a Plastic Substrate. *IEEE Electron Device Lett.* **2013,** *34* (2), 160–162. DOI: 10.1109/LED.2012.2228464.
21. (10) (PDF) Fast Flexible Transistors with a Nanotrench Structure. ResearchGate. https://www.researchgate.net/publication/301568917_Fast_Flexible_Transistors_with_a_Nanotrench_Structure (accessed on 09 Apr 2020).
22. (10) (PDF) All-Printed Diode Operating at 1.6 GHz. ResearchGate. https://www.researchgate.net/publication/264643475_All-printed_diode_operating_at_16_GHz (accessed on 09 Apr 2020).

23. (10) (PDF) High- Mobility Polymer Gate Dielectric Pentacene Thin Film Transistors. ResearchGate. https://www.researchgate.net/publication/228554325_High- Mobility_Polymer_Gate_Dielectric_Pentacene_Thin_Film_Transistors (accessed 09 Apr 2020).
24. Conjugated Polymer—An Overview | ScienceDirect Topics. https://www.sciencedirect.com/topics/materials-science/conjugated-polymer (accessed on 10 Apr 2020).
25. Zhu, W. et al. Graphene Radio Frequency Devices on Flexible Substrate. *Appl. Phys. Lett.* **2013,** *102* (23), 233102. DOI: 10.1063/1.4810008.
26. Lin, C.-Y. et al. High-Frequency Polymer Diode Rectifiers for Flexible Wireless Power-Transmission Sheets. *Org. Electron.* **2011,** *12* (11), 1777–1782. DOI: 10.1016/j.orgel.2011.07.006.
27. Power, A. C.; Gorey, B.; Chandra, S.; Chapman, J. Carbon Nanomaterials and Their Application to Electrochemical Sensors: A Review. *Nanotechnol. Rev.* **2018,** *7* (1), 19–41. DOI: 10.1515/ntrev-2017-0160.
28. (10) (PDF) Nanotube Electronics for Radiofrequency Applications. ResearchGate. https://www.researchgate.net/publication/40039419_Nanotube_electronics_for_radiofrequency_applications (accessed on 10 Apr 2020).
29. Bolotin, K. I. et al. Ultrahigh Electron Mobility in Suspended Graphene. *Solid State Commun.* **2008,** *146* (9–10), 351–355. Doi: 10.1016/j.ssc.2008.02.024.
30. Kaur, A.; Yang, X.; Park, K. Y.; Chahal, P. Reduced Graphene Oxide Based Schottky Diode on Flex Substrate for Microwave Circuit Applications. In *2013 IEEE 63rd Electronic Components and Technology Conference*; **2013**; pp 1037–1042. DOI: 10.1109/ECTC.2013.6575700.
31. Ozaki, H.; Kawamura, T.; Wakana, H.; Yamazoe, T.; Uchiyama, H. 20-μW Operation of an a-IGZO TFT-Based RFID Chip Using Purely NMOS 'Active' Load Logic Gates with Ultra-Low-Consumption Power. In *2011 Symposium on VLSI Circuits—Digest of Technical Papers* **2011**; pp 54–55.
32. Halonen, E.; Kaija, K.; Mantysalo, M.; Kemppainen, A.; Osterbacka, R.; Bjorklund, N. Evaluation of Printed Electronics Manufacturing Line with Sensor Platform Application. In: *2009 European Microelectronics and Packaging Conference*; **2009**; pp 1–8.
33. Castro, A. T.; Sharma, S. K. Inkjet-Printed Wideband Circularly Polarized Microstrip Patch Array Antenna on a PET Film Flexible Substrate Material. *IEEE Antennas Wirel. Propag. Lett.* **2018,** *17* (1), 176–179. DOI: 10.1109/LAWP.2017.2779440.
34. Ramadan, M.; Dahle, R. 3-D Printed Flexible Heterogeneous Substrates with Customizable Gain and Bandwidth. In: *2018 18th International Symposium on Antenna Technology and Applied Electromagnetics (ANTEM)*; **2018**; pp 1–4. DOI: 10.1109/ANTEM.2018.8572880.
35. Moussessian, A. et al. An Active Membrane Phased Array Radar. In: *IEEE MTT-S International Microwave Symposium Digest* **2005,** *2005*, 4. DOI: 10.1109/MWSYM.2005.1517047.
36. Alharbi, S.; Kiourti, A. Folding-Dependent vs. Folding-Independent Flexible Antennas on E-Textiles. In: *2019 IEEE International Symposium on Antennas and Propagation and USNC-URSI Radio Science Meeting*; **2019**; pp 755–756. DOI: 10.1109/APUSNCURSINRSM.2019.8888646.

CHAPTER 7

Wide Band Low-Profile High Directive Antennas for 5G Applications

SASMITA MOHAPATRA

*Department of Electronics and Communication Engineering,
Sir M. Visvesvaraya Institute of Technology, Bengaluru,
Visvesvaraya Technological University, India*

ABSTRACT

Technology has always its way out of satisfying people in achieving all comforts for which fifth generation (5G) technology is proven to be a boon for complete mankind. The intensive research work carried out to make 5G communication useful for many applications shows the innovation to meet the growing demand for high-speed wireless communications. At present, 5G technology has many applications which not only include high-speed mobile networks but also it has applications related to Internet of Things (IoT) that help modernize homes, cities, and farming and make all smart. Antenna configuration is a basic requirement in any wireless communication as the efficiency and coverage of the communication system wholely depend on the type of antenna used. This paper suggests an efficient antenna that consists of a multiple input multiple output (MIMO) structure with Antipodal Vivaldi antenna which is very compact in structure with very wide range of coverage and high directivity. Thus many performance enhancement techniques have been maneuvered in designing the antenna including choosing a correct substrate, corrugation process, multi-element technology, mutual coupling reduction techniques, etc.

5G Green Communication Networks for Smart Cities. Devasis Pradhan, Rajeswari, Hla Myo Tun, Naw Khu Say Wah, & Thandar Oo (Eds.)

7.1 INTRODUCTION

5G technology has been considered as a future-generation socket that does not only include the mobile communication rather will power a vast range of industries starting from retail to education, smart homes to healthcare, and transportation to entertainment. It is the following era technology that will furnish high data rates, faster communication, wide capacity, and a top quality of service. 5G technology will definitely unleash a variety of new ways to break through conventional obstacles to development. 5G technology encourages the IoT which gives major social changes in the field of education, farming, industry, healthcare, security and surveillance, and different spheres of society. 5G technology has extensive applications supporting IoT which is useful in logistics and shipping, fleet management, vehicle automation, drone operation, etc. As such, 5G technology will allow more devices to be connected, and the network will be able to meet the communication needs while maintaining a compromise between delay, speed, and price.

5G technology has vast applications in mainly three areas which include

- Enhanced mobile broadband—5G Technology makes smartphones better. 5G cellular technologies usher in brand-new immersive media like VR and AR, delivering faster and more consistent data rates, decreased latency, and decreased bit cost.
- Mission-crucial communication via 5G allows new offerings that could remodel industries with extremely dependable, accessible, and low-latency systems hyperlinks like far-flung manipulation of crucial infrastructure, vehicles, and clinical procedures.
- The massive IoT-5G is supposed to flawless join a large range of virtually implanted sensors the entirety through the capacity to diminish in data rates, power, and mobility—imparting extraordinarily lean and inexpensive connectivity solutions.

There are mainly 15 specifications associated with 5G technology.

As per the above figure, 5G communication is steered by fifteen required provisions namely Deployment, data bandwidth, technology, services, multiplexing, switching, core network, frequency bands, latency, access technology, spectrum efficiency, data rate, connection density, forward error correction code, and mobility considering the connection's reliability.[3–5]

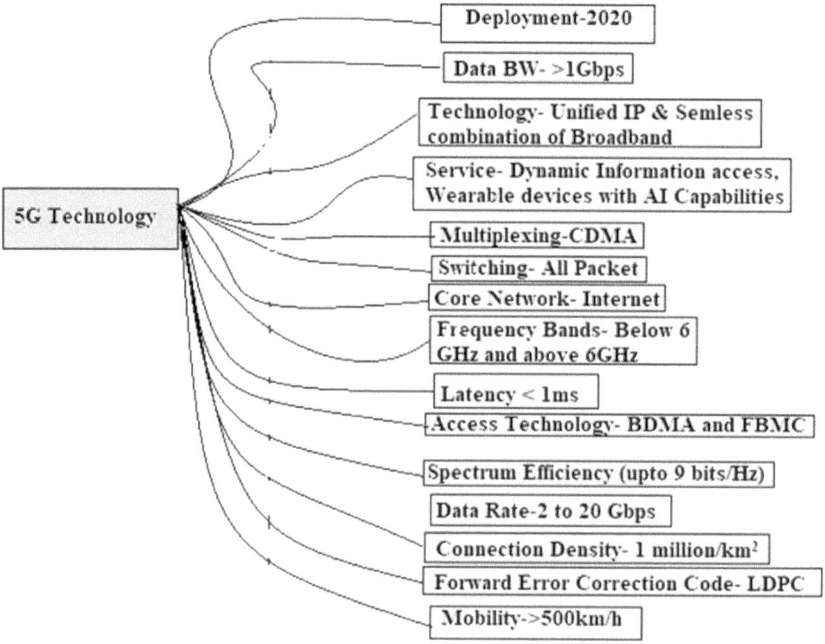

FIGURE 7.1 Provisions of 5G technology.

5G can be much faster than 4G which uses millimeter waves and can be functional both in nether bands (e.g., sub 6 GHz) and higher bands (greater than 6 GHz). It can provide faster data rates up to 20 Gbps and 100 plus Mbps mean data rates. To gain low latency, low density parity checks, codes are utilized as a forward error correcting code. The 5G generation supports filter bank multicarrier (FBMC) and beam division multiple access (BDMA). Multiple space division access termed as BDMA can be achieved because of highly directive beams of radio transmission. Better spectral efficiency can be achieved by the use of a bank of filters in FBMC. Because of the low latency feature, 5G technology can be used for real-time video of exceptional quality.

The heart of any wireless communication system is the antenna system used. Thus for the efficient working of any 5G system the antenna used must be very efficient with respect to its directivity, bandwidth, coverage, and lowest loss with low profile construction. Thus the required antenna design for 5G communication becomes very crucial which should satisfy all the required characteristics to make the system very efficient.

Considering the above-mentioned characteristics in this research paper the best suitable antenna has been designed and suggested which is chosen to be MIMO type with respect to input-output ports containing an array of antipodal Vivaldi antenna with respect to antenna type. For enhancing the performance of the antenna, special care has been taken while choosing the substrate for the construction of the antenna and for reducing the mutual coupling between the antennas electromagnetic bandgap (EBG) Structure has been used. Here to increase the multi-element antenna's gain and transmission range structure without any metal rim is been used.

7.2 MAIN FEATURES OF ANTENNA

According to Refs. [1, 2], an array-based UWB microwave antenna system is very useful for RADAR System. In Ref. [1], an array of Antipodal Vivaldi antennas have been considered for efficient use for the RADAR System. The same concept has been carried out in the present work where for an efficient working of the array antenna first individual antenna has been given importance. Here, also the individual antenna used are Antipodal Vivaldi antenna used for wider coverage with low loss. The antenna chosen here is a MIMO antenna system containing all Antipodal Vivaldi antenna arranged in an array structure. The main characteristics of the antenna system are: MIMO Structure Antipodal Vivaldi antenna RO4003 dielectric material EBG Structure Multi-element antenna without any metal rim.

7.2.1 MIMO STRUCTURE

MIMO structure is mainly used to enhance the range of maximum transmission without increasing any signal power. Wireless communication system has some challenges like interference, radiation losses, multipath fading, and long-range communication. All the challenges are to be addressed at a very high frequency when it is considered for 5G Technology. MIMO system is best suitable for any 5G system which can bring large efficiency to the system by reducing latency and achieving very high throughput. In any MIMO System, the channel capacity can be improved significantly by the use of multiple antennas. Also, care can be taken to decrease the number of antennas in any MIMO System by using multiband antennas that supply the reach of various wireless applications.[6] Depending on the frequency bands these MIMO antennas are categorized as broadband as well as multiband antennas.

These broadband and multiband antennas can be structured according to their applications as multiband antennas have metal rims and multiband antennas do not have metal rims. The multi-element antenna with a metal rim provides excellent mechanical strength. A MIMO antenna with high transmission speed is desirable where maximum isolation is incorporated. In order to improve the gain while establishing maximum isolation between antennas different types of enhancements are done in the antenna structure. These techniques also affect bandwidth, efficiency, and envelope correlation coefficient (ECC). The electromagnetic exchange of antenna components is known as mutual coupling. İf mutual coupling is not addressed properly in any MIMO System, energy is consumed when one antenna is receiving and another is still emitting. Thus mutual coupling reduction between the antenna elements is very much essential which is calculated as

$$MC_{mn} = exp\left(-\frac{2x_{mn}}{\lambda}(\alpha + n\pi)\right), \quad m \neq n \tag{7.1}$$

$$MC_{mn} = 1 - \frac{1}{N}\sum_{m}\sum_{m \neq n} MC_{mn} \tag{7.2}$$

Where MC_{mn} = mutual coupling between mth and nth antenna elements

x_{mn} = distance between m^{th} and n^{th} antenna elements

In the above equation

α = coupling level control

N = number of MIMO elements

The other important parameter associated with MIMO Structure is the ECC that specifies correlation of the signals coming in on a MIMO port. ECC can be determined by using the formula as per[7]:

$$\left|p_{mn}(e)\right|^2 = 1 - \frac{\eta_{max}}{\eta_m}\frac{1}{\eta_n} \tag{7.3}$$

Where

$p_{mn}(e)$ = correction coefficient between mth and nth port

η_{max} = Maximum efficiency

$\eta_m \eta_n$ = Total efficiency offered by all the radiating elements

7.2.2 ANTIPODAL VIVALDI ANTENNA

Vivaldi antenna is a special type of antenna that is associated with high gain and wider frequency band operation. It mainly comprises two conductors on either side of the board which are mirror images to one another. Usually, a radiator is framed by the upper conductor and a ground plane is framed by the bottom conductor.[8–11] A tapered slot is present with each of the antipodal Vivaldi antennas which has been inscribed on a metal film of very minimum thickness. Mostly the Antipodal Vivaldi antenna operates as a high directive end-fire antenna with a superior directivity associated with broadband operation.

7.2.3 RO4003 Dielectric Material

Rogers RO4003 is a High-Frequency Circuit Material that is fortified hydrocarbon/ceramic laminates and designed for sensitivity in performance, and a lot of commercial applications. This dielectric retains a dielectric constant of 3.38 and a loss tangent of 0.00027. This dielectric is designed to provide high-frequency performance and low-cost fabrication of circuits. Thus it results in a low loss material. Usually, it comes with a thickness of 0.508 mm with a size of (40×60) mm^2. The gain of the antenna using this substrate can be achieved up to (8–15) dB and the working frequency ranges from 3.4 to 40 GHz.

7.2.4 EBG STRUCTURE

EBG structure usually works as a transmission medium for electromagnetic waves. Usually, this gap has a periodic structure which is made up of dielectric or metal material. Thus this can produce more than one band gap due to periodicity independence resonance. Usually, this EBG structure is used in any antenna system to achieve low mutual coupling and high efficiency.[7]

7.2.5 MULTI-ELEMENT ANTENNAS

In the 5G communication system, the used antennas are expected to have certain characteristics like high gain, ultra-wide frequency band, and stable radiation pattern. It is difficult to have all the above characteristics in a single antenna. So multi-element antenna structure is preferred for 5G communication system.[8,12]

7.2.6 MULTI-ELEMENT ANTENNA WITHOUT ANY METAL RIM

Usually, the overall signal amplification and transmission range are improved by the MIMO multi-element antenna. This is a very basic requirement in any communication system. Multi-element antennas can be with rim and without rim. Mainly MIMO multi-element structure is used in 5G System to improve signal gain and transmission range. These characteristics are a very basic and important requirement for any high-speed communication system. MIMO element antennas can be slot antennas, differently structured monopoles dual and tri polarized antennas. Together with MIMO technology cognitive radio (CR) is also taken as a core technology for 5G System. An amalgamation of CR and MIMO creates antennas with more efficiency where the radiation efficiency and data rate are improved for the antenna system. MIMO challenge is related to mutual coupling between the antennas. Mutual coupling is reduced in this case by using an EBG structure.

7.3 INDIVIDUAL ANTIPODAL VIVALDI ANTENNA STRUCTURE DESIGN AND ANALYSIS

The said antenna for 5G System is a MIMO structure where individual antipodal Vivaldi antennas are ordered in a N×N matrix structure. In the said antenna structure individual antipodal Vivaldi antennas consists of tapered slot which has been inscribed on a metal film of very low thickness. The dielectric used here is Rogers RO4003 due to its electrical characteristics and high stability. The antenna is preferred because of its ultra gain, broad bandwidth, minimal radiation loss, and stable radiation pattern. Mainly the antenna behaves as an end-fire antenna with very high directivity. So the said antipodal Vivaldi antenna is being applied in many applications to have an ultrawideband application with wider coverage.

The losses that occur by any antenna mainly depend on the feeding structure. Impedance matching at the feeding end is a big challenge for any antenna which decides the radiation loss of the antenna. The said Antipodal

Vivaldi antenna has been structured to have a balun providing matching conditions between the antenna and 50 Ω microstrip feed line. The balun comprises of a 50 Ω microstrip-to-parallel-plate transmission line tapering to a slotline arrangement, and the said antenna is excited through the slot line structure. In this present research carried, an antipodal Vivaldi antenna with a low profile tapered slot has been depicted. Ansys HFSS simulation has been used to design the said antipodal Vivaldi antenna and its different characteristics have been analyzed. The antenna's basic execution with regard to several parameters is found to be very tremendous over a broad frequency range of 2.0 to 20 GHz.

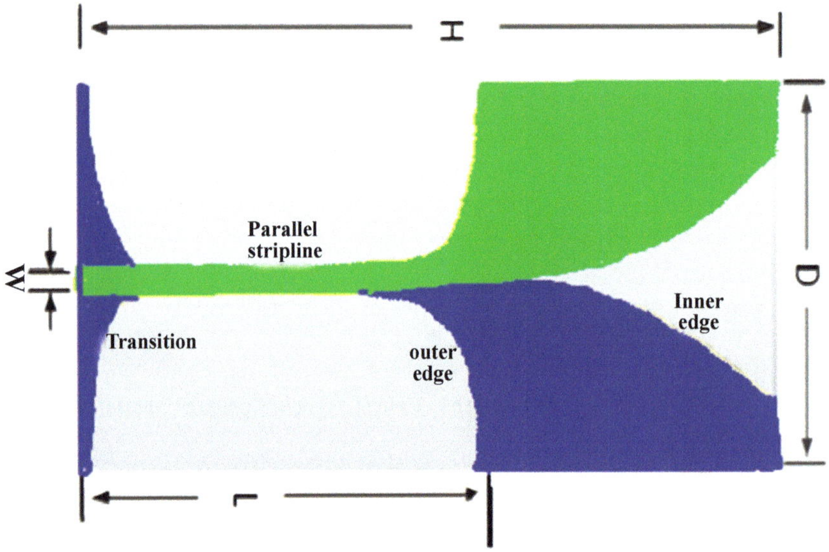

FIGURE 7.2 Prototype diagram of the antipodal Vivaldi antenna.

The above figure depicts the configuration of the proposed Antipodal Vivaldi antenna. The main focus in constructing the antenna is with respect to three major parts which play a very important part in the manufacturing of the antenna. The first part depicts the passing of signal from the microstrip line to the configured parallel stripline with two sides that takes place in the area of tapered part flared circularly. The next part is the major carrier which carries the signal from the transmission line to antenna at the exciting end that is the parallel stripline. This is a balanced structure where transitions take place over a broad frequency range. The third and major a portion of the overall construction is the flaring radiation zone which is responsible for converting the bounded wave to a unbounded wave or radiation. This radiating zone is mainly depicted by exponential equations that define curves, which incorporate the inner and outer edges.

To reduce the reflections due to impedance mismatch the external lines are tapered consisting of ¼ elliptical arcs. The equations for designing the individual antipodal antennas are stated as,

$$V = C_1 e^{\alpha u} + C_2 \tag{7.4}$$

$$\text{Where } C_1 = \frac{V_2 - V_1}{e^{\alpha u_2} - e^{\alpha u_1}} \tag{7.5}$$

$$\text{and } C_2 = \frac{V_1 e^{\alpha u_2} - V_2 e^{\alpha u_1}}{e^{\alpha u_2} - e^{\alpha u_1}} \tag{7.6}$$

Where α = the tapered line's axis ratio

(u_1, v_1) = starting point of the taper line

(u_2, v_2) = ending point of the taper line

The transition region where from microstrip to parallel strip line and parallel strip line to slot line take place is more challenging in designing. During high-frequency operation, these transition regions can lead to discontinuities so they must be designed very smoothly. As the designed antenna is to be placed in array arrangement for MIMO antenna so lot of supervision has been taken for reducing the size of antenna. So the designed antenna is a low profile antenna.

According to the required specification and application, the antenna is designed and well optimized with a dimension of 50 × 20 mm. The various depicted parameters are w = 3.00mm, L = 20mm, ε = 3.38, and h = 0.058 mm.

Ansys HFSS software has been used and the results found after simulation for the Return Loss for the individual Antipodal Vivaldi antenna are shown in Figure 7.3. It is observed that the Return Loss is nearer to − 10 dB from 2.0 to 20GHz, which signifies that the performance of the antenna is steady across a broad frequency band.

The 3D pattern for radiation of each antenna has been examined for different frequencies. The performance of the antenna is found to be stable over a broad band of frequencies.

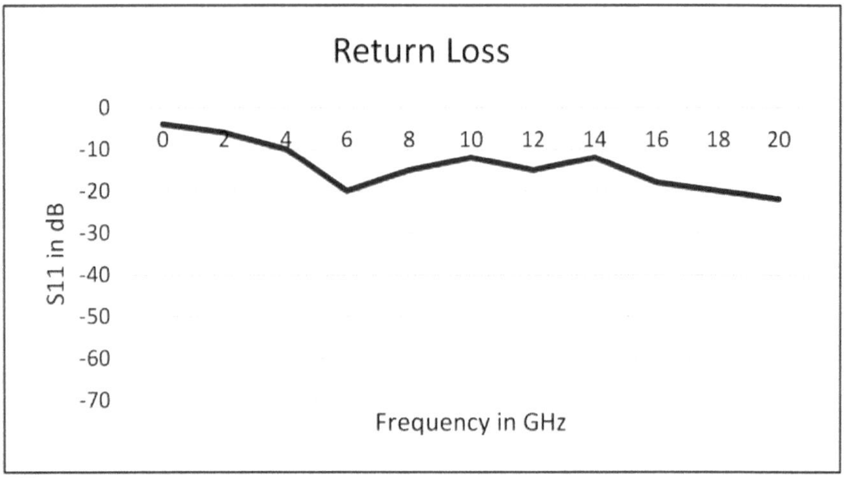

FIGURE 7.3 Return Loss for individual antenna.

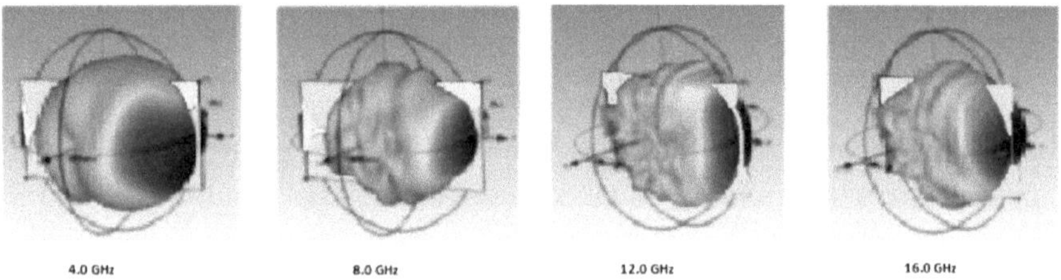

FIGURE 7.4 3D pattern for radiation.

The above figure shows the result after simulation for the calculation of gain for a single Antipodal Vivaldi antenna, indicating that the said antenna keeps a good gain across the whole frequency range. The antenna gain is found to be around 7.0 dB at 10 GHz.

7.4 MIMO STRUCTURE WITH ARRAY OF ANTIPODAL VIVALDI ANTENNA

Whenever in a wireless system wide coverage is required, an array of antennas plays a vital role. This array of antenna also helps target the destination by increasing the signal strength thus increasing the overall directivity of the system. Usually by using array of antennas radiation loss can be reduced by removing the minor lobes and making the radiation unidirectional improving the signal-to-noise ratio. However, based on how many antennas are included in the array structure and how they are arranged in the array, the design can generate a very big gain overall. Thus, the accomplishment of the overall system can be improved by using an array structure where power wastage is reduced.

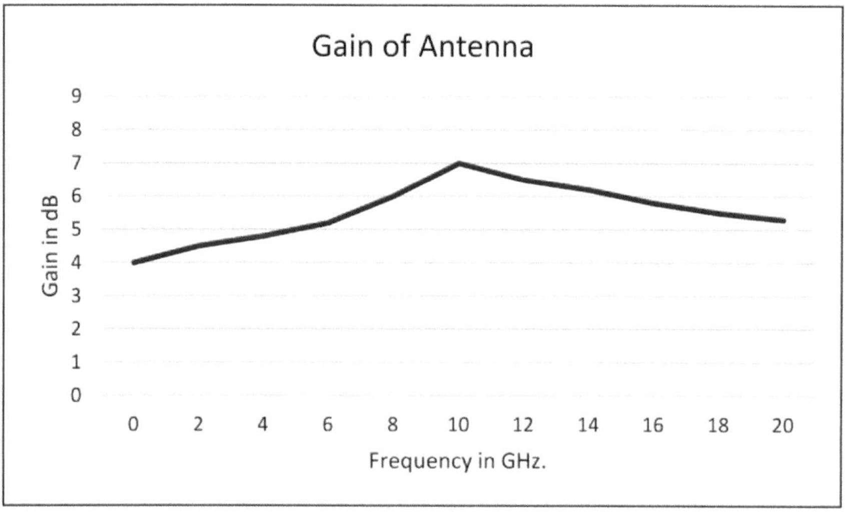

FIGURE 7.5 Simulated gain.

In the present MIMO structure, the antipodal Vivaldi antennas are ordered in an N × N array fashion. In the present case, a 4 × 4 array has been designed.

As discussed in Ref. [1] designing an array antenna has many challenges. The main challenge arises due to feeding of the antenna as while feeding maximum care has to be taken for distributing the input equally between all the antennas. This distribution decides a lot regarding the antenna gain. One biggest challenge is the cross-coupling between the antennas. To reduce cross coupling the side lobes of the antennas must be reduced. For this reason, care has to be taken to improve the gain of the individual antenna. The antennas are here fed using corporate feed and series feed techniques for excitation. EBG structure has been used which has a periodic structure. This produces more than one band gap due to periodicity independence resonance. This EBG structure is used in antenna systems to achieve low mutual coupling and high efficiency. The antenna is designed without any metal rim structure.

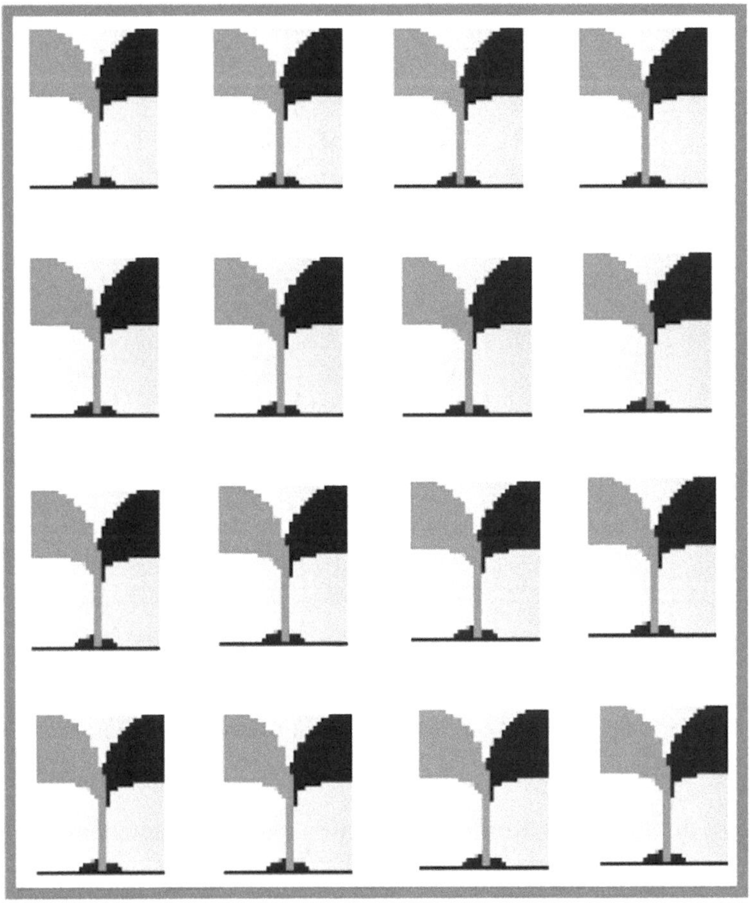

FIGURE 7.6 Multi-array antenna design.

In this section, the antenna array has been analyzed with respect to its VSWR and Return Loss.

The found results for the array of antenna in 4 × 4 matrix show that the antenna array has a very good VSWR which varies within a range from 3 to 2.8 over a wide band of operating frequency of 2 to 16 GHz. This shows the impedance matching of the structure. Thus this proves that the distribution of the input signal over the antenna array is good.

Similarly, the array of antennas is analyzed with respect to Return Loss.

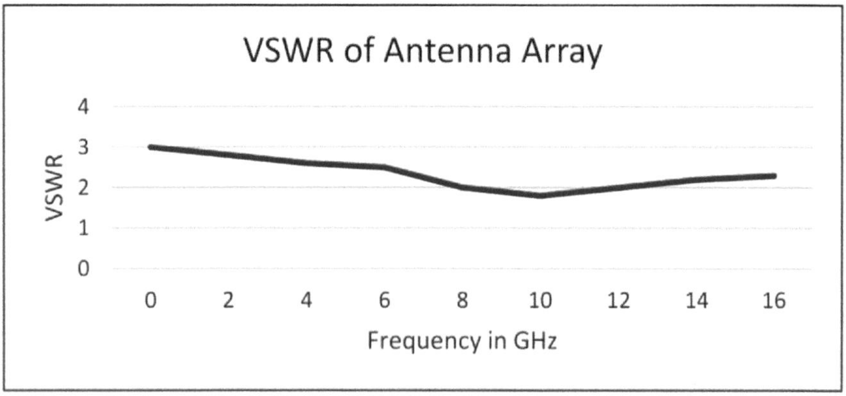

FIGURE 7.7 Simulated result of VSWR for the designed antenna array.

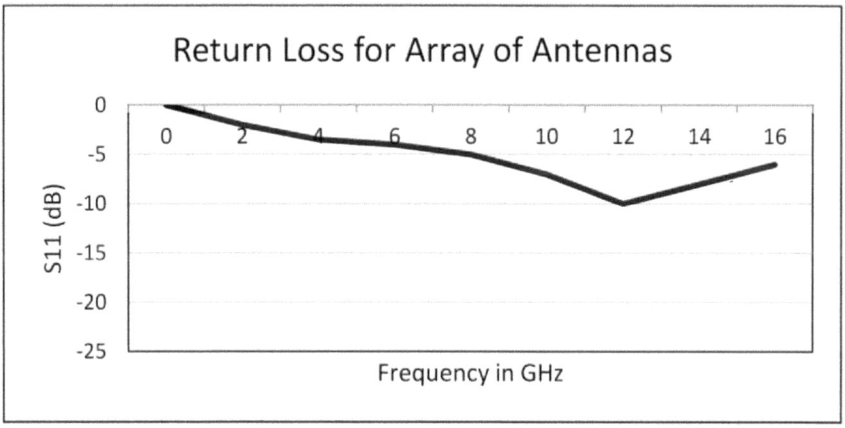

FIGURE 7.8 Return Loss for the designed antenna array.

Figure 7.8 shows the analysis of Return Loss for the antenna array. The Return Loss over the wide range from 2.0 to 16.0 GHz is almost constant within −10 dB. This analysis shows that the gain of each antenna is good with less individual loss.

7.5 CONCLUSION AND FUTURE SCOPE

The said array of antenna has been designed and simulated for the analysis over a broad band of frequencies from 2.0 to 20.0 GHz with respect to various parameters like VSWR and Return Loss of antenna. The results are found to be very suitable with low loss and good impedance matching which is ideal for any communication center for wide coverage. Thus, this MIMO antenna can be preferred in the case of a 5G communication system.

Anyways similar array antennas with different options of individual antenna can be designed and analyzed to find the best suitable antenna for the 5G system. The other parameters of the antenna can be analyzed over a wide range of frequencies to check the efficiency of antenna.

KEYWORDS

- **5G technology**
- **MIMO system**
- **Vivaldi antenna**
- **array of antennas**
- **high directive**
- **low profile**
- **higher operating range**

REFERENCES

1. Mohapatra, S. Antipodal Vivaldi Antennas Arranged in Circular Array for RADAR. In: *Smart Antennas. EAI/Springer Innovations in Communication and Computing*; Malik, P. K., Lu, J., Madhav, B. T. P., Kalkhambkar, G., Amit, S., Eds.; Springer: Cham, **2022**.
2. Mohapatra, S. Array of Vivaldi antenna arranged in Circular Fashion for MTI RADAR. *J. Xidian Univ.* **2020,** *14* (8), 685–693.
3. Huang, H.-C. Overview of Antenna Designs and Considerations in 5G Cellular Phones. In: *2018 International Workshop on Antenna Technology (iWAT)*, **2018**; pp 1–4.
4. Hong, W. et al. Multibeam Antenna Technologies for 5G Wireless Communications. *IEEE Trans. Antennas Propagation* **2017,** *65* (12), 6231–6249.

5. Guidelines for Evaluation of Radio Interface Technologies for IMT-2020, Document ITU-R M.2412-0, **2017**; pp. 1–144.
6. Khan, R.; Al-Hadi, A. A.; Soh, P. J.; Kamarudin, M. R.; Ali, M. T.; Owais. User Influence on Mobile Terminal Antennas: A Review of Challenges and Potential Solution for 5G Antennas. *IEEE Access* **2018,** *6*, 77695–77715.
7. Nadeem, I.; Choi, D.-Y. Study on Mutual Coupling Reduction Technique for MIMO Antennas. *IEEE Access* **2019,** *7*, 563–586.
8. Dixit, A. S.; Kumar, S. A Miniaturized Antipodal Vivaldi antenna for 5G Communication Applications. In: *2020 7th International Conference on Signal Processing and Integrated Networks (SPIN)*; **2020**; pp 800–803.
9. Goel, T.; Patnaik, A. Novel Broadband Antennas for Future Mobile Communications. In: *IEEE Trans. Antennas Propagation* **2018,** *66* (5), 2299–2308. DOI: 10.1109/TAP.2018.2816660.
10. Dixit, A. S.; Kumar, S. A Survey of Performance Enhancement Techniques of Antipodal Vivaldi antenna. *IEEE Access* **2020,** *8*, 45774–45796.
11. Dixit, A. S.; Kumar, S. The Enhanced Gain and Cost-Effective Antipodal Vivaldi Antenna for 5G Communication Applications. *Microw. Opt. Technol. Lett.* **2020,** *62* (6), 2365–2374.
12. Bhadoria, B.; Kumar, S. A Novel Omnidirectional Triangular Patch Antenna Array Using Dolph Chebyshev Current Distribution for C-Band Applications. *Progress Electromagn. Res.* **2018,** *71*, 75–84.

CHAPTER 8

Table Lamp Inspired Miniaturized Fractal Antenna for 5G Enabled Green Communications in Smart Cities

ATUL VARSHNEY[1], VIPUL SHARMA[1], T. MARY NEEBHA[2], and N. PRASANTHI KUMARI[3]

[1]*ECE Department, FET, Gurukula Kangri (Deemed to be University), Haridwar, Uttarakhand, India*

[2]*ECE Department, Karunya Institute of Technology and Science, Coimbatore, Tamilnadu, India*

[3]*Electrical Department, SoE, UPES, Dehradun, Uttarakhand, India*

ABSTRACT

This chapter presents a low-cost transformer-fed miniaturized monopole based on a circular fractal antenna for 5 G-enabled smart cities Sub-6GHz (n78 band) short-range green communications. An 85.33% miniaturized size of the antenna is achieved by the two-level circular fractal technique and compressed partial ground plane (CPGP) methodology. The miniaturized antenna is fabricated on a low-profile FR4 substrate and tested using vector network analyzer MODEL N9917A. A transformer fed is utilized for better impedance matching and good transfer of RF energy. The measured reflection coefficient and voltage standing wave ratio (VSWR) are found in excellent agreement. The proposed antenna has a peak gain of 2.88 dBi and a radiation efficiency of approximately 96% at 3.5GHz. The designed antenna has a bandwidth of 3.23GHz to 3.80GHz. Thus the proposed antenna

5G Green Communication Networks for Smart Cities. Devasis Pradhan, Rajeswari, Hla Myo Tun, Naw Khu Say Wah, & Thandar Oo (Eds.)

is a good candidate for WiMax (3.3–3.4GHz/3.4–3.6GHz/3.6–3.8GHz) and Sub6GHz (3.3–3.8GHz) 5G applications for smart cities.

8.1 INTRODUCTION

As per policies of different governments to construct smart cities all over the world a number of devices, sensors, and embedded platforms are deployed across our surrounding environment to make this dream a reality. Therefore, increased demands of various systems and sub-systems in smart cities day by day are projected to get better for our comfort. This may demand energy requirements to connect billions of new devices and trillions of newly generated message exchanges among these devices that would add significant emissions of greenhouse gas. Therefore for sustainable, smart cities green communication enabling technologies and their solutions are critical.[1,2] In smart cities, green communication is challenging because of the traffic necessities and deployment scenarios, a large collection of hardware and equipment concerned, a huge machine-type device, a miscellaneous range of applications with unlike qualities of services, tradeoffs between numerous energy efficiency, and mixed communication technologies and networking protocols and architectures requirements.[1] Machine-to-machine (M2M) communication techniques smooth the progress of nonstop communication between cellular users. Reduction in the backhaul communication necessities and latency is possible with a huge number of small-cell techniques in 5G communications. An efficient novel algorithm for M2M communication is proposed in Ref. [3]. 5G communication systems have an extensive choice of frequency bands. Y. Wei and S.H. Hwang have investigated maximizing the efficiency in terms of cost, spectrum, and energy for green communication in a smart city using the concept of radio spectrum value, the selection of appropriate frequency bands in a different line of sight, non-line of sight, indoor and outdoor.[4] A. Fatima et al. have proposed algorithms to save energy and costs for multiple virtual machine placement via Bin packing in cloud data centers for load balancing problems in cloud data centers with unbalanced network resources.[5] C. Lork et al. have proposed an ontology-based framework for building energy management using IoT.[6] M. Zahid et al. have proposed a traffic flow splitting model for 5G vehicle networking terminals that jointly optimizes delay and cost and provides an efficient splitting of services in 5G vehicle-to-X (V2X) networks.[7]

Antenna plays a vital role in 5G enabling cities for green communication systems for connectivity between billions of devices and message transfer. For 5G communication, most countries like China, the USA, Europe, and India use the n78 (3300-3800MHz) band. For green communication low-profile, low gain high radiation efficiency antennas are required for short-range communications using Wi-Fi, Bluetooth, WLAN, IoT, WiMax, and 5G frequency bands (Sub-6GHz). 6Funda Cirik and Bahadir Sugleyman Yildirim 2014 have proposed a very high gain narrow band rectangular patch antenna for WiMAX and 5G applications. The biggest drawback of this antenna is its big size and 3D structure.[8] Dhatu Paragya et al., 2020 have demonstrated proximity coupled, linearly polarized, high gain, narrow band, microstrip antenna for n78 5G mobile communications.[9] Ikhlas Ahmad et al., 2021 have proposed a 12-PIN diode-based frequency reconfigurable monopole antenna for Sub-6GHz (n78 band) 5G communications. The antenna has low gain, and a narrow bandwidth in five operating bands.[3,10] Nayla Ferdous et al. (2019) have proposed offset edge fed, monopole, and compact elliptic patch antenna with high gain narrow band for 5G and WiMAX applications.[5,11] An. Wenxing et al. (2018) have demonstrated a hybrid dual dipole edge feed antenna with high gain and wideband performance for 5G mobile communications. The drawback of this antenna is its 4.5mm thickness.[12]

In this chapter, the main objective is to design and fabricate an antenna with coverage of full n78 band (3300-3800MHz) with a low profile, medium gain, and high radiation efficiency for WiMax and 5G communications systems. Another objective is to miniaturize the antenna size up to a large extent without much effect on its wide bandwidth for coverage of n78 5G signals. So that it can easily be fixed with any of the mobile, IoT devices and systems to facilitate green communication in smart cities.

8.2 DESIGN DEVELOPMENT AND OPTIMIZATION

8.2.1 ANTENNA DESIGN EQUATIONS

The antenna is derived from circular patch fractal techniques. Thus these fundamental principle design equations (8.1) to equation (8.4)[13,14] were used for the analysis of the antenna:

Radius of Circular Patch

$$a = \frac{F}{\{1 + \frac{2h}{\pi \varepsilon_r F}[\ln\left(\frac{\pi F}{2h}\right) + 1.7726]\}^{1/2}} (\text{cm}) \quad (8.1)$$

Where,

$F = \frac{8.791 \text{ X}10^9}{f_0 \sqrt{\varepsilon_r}}$, ($f_0$ in Hz and h must be in cm)

Length of the 50 Ω Feed line

$$F_i = 10^{-4}[6697 - 4043\varepsilon_r + 2561.9\varepsilon_r^2 - 682.69\varepsilon_r^3 + 93.187\varepsilon_r^4 - 6.1783\varepsilon_r^5 + 01376\varepsilon_r^6 - 0.001699\varepsilon_r^7]\frac{\text{Lsub}}{2} \quad (8.2)$$

Substrate and ground width and length

$$w_g = l_g = 2(\text{diameter}) = 2(2a) = 4a \quad (8.3)$$

50 Ω Microstrip Feed width

$$W_{ms} = \frac{7.48h}{e^{(Z_0 \frac{\sqrt{\varepsilon_r + 1}}{87})}} \quad (8.4)$$

Quarter Wave Impedance Transformer Impedance

$$Z_{T/F} = \sqrt{(Z_0.Z_{Patch})} \quad (8.5)$$

Generally, patch impedance lies between 100 and 400 Ω. Normally, edge impedance of a circular patch is 243 Ω. Impedance transformer width is also calculated using equation (8.4) corresponding to the impedance of quarter wave impedance transformer using equation (8.5).

8.2.2 ANTENNA CONFIGURATION AND DESIGN DEVELOPMENT

The proposed antenna is designed on a chosen FR4-epoxy substrate. The thickness of the substrate is 1.6mm with a loss tangent of 0.02. The antenna is designed at an operating frequency 3.5GHz. It is fabricated on a piece of 13mm x 26mm substrate. Two levels of fractal iterations were used to achieve the table lamp-shaped antenna as shown in Figure 8.1(a).

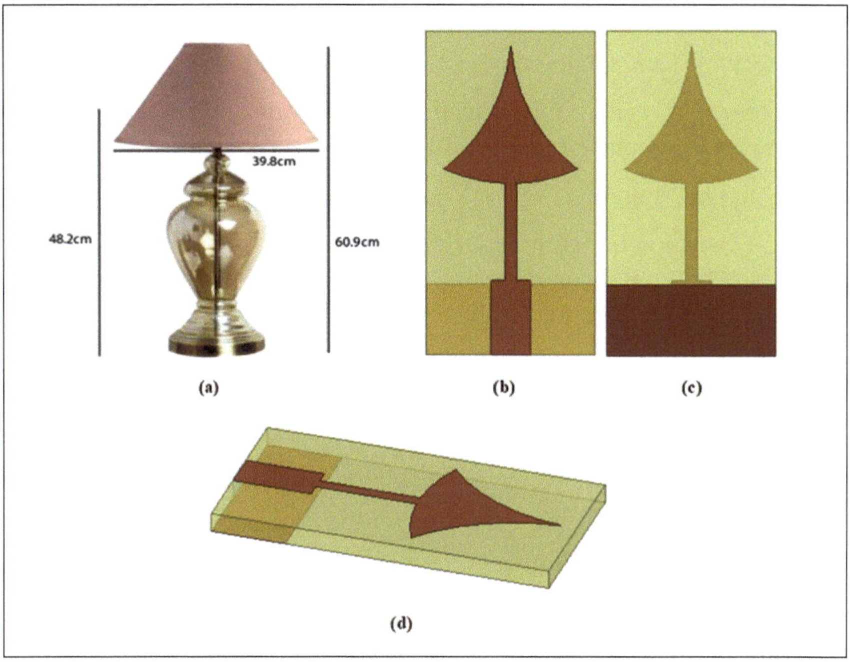

FIGURE 8.1 Antenna configuration: (a) table lamp,[15] (b) top view, (c) rear view, and (d) trimetric view.

The designed antenna configuration top, rear, and trimetric views are represented in Figure 8.1(b–d). All fractal operation stages on patch are illustrated in Figure 8.2(a). Three circles are used for fractals of patch to get a targeted table lamp-shaped antenna. Initially in iteration zero ('0') the circular patch antenna is designed using equations (8.1–8.3). The radius of the fundamental circular patch using equation (1) is 11.51mm centered at 25.775mm from the RF input port. Its optimized value is 12mm ('a') for minimum reflection coefficients and the basic size of antenna using equation (3) is 48mm x 48mm as displayed in Figure 8.2(b). A 3mm wide 50Ω feed line is used for excitation and an impedance transformer is applied between patch and 50Ω feed line for maximum RF energy transfer. The full ground surface at the bottom layer of the substrate results in a narrow bandwidth. Therefore, to improve antenna performance for wideband applications the length of the antenna is reduced to 5.7mm. Thus the unidirectional radiation pattern becomes bidirectional. This ground reduction (partial ground structure) results in a monopole patch antenna. In iteration first ('1') a circular cut of radius, 7 mm $(a_3 = (\frac{a}{2}+1)\ \text{mm})$ centered at 31.75 mm ('C_1') on vertical bisection line from the input excitation end is subtracted as displayed in Figure 8.2(c). In iteration first ('1') a circular cut of radius, 7 mm $(a_3 = (\frac{a}{2}+1)\ \text{mm})$ centered at 31.75mm ('C_1') on the vertical bisection line from input excitation end is subtracted as displayed in Figure 8.2(c). In iteration second ('2) two circles of radii, a_1=a_2=2a_3=14 mm each centered symmetrically along horizontal direction at a distance ± 14mm, and at 25.775 mm on vertical bisection line from the input excitation end are subtracted from iteration first ('1'), this yields the desired table lamp-shaped patch as depicted in Figure 8.2(d).

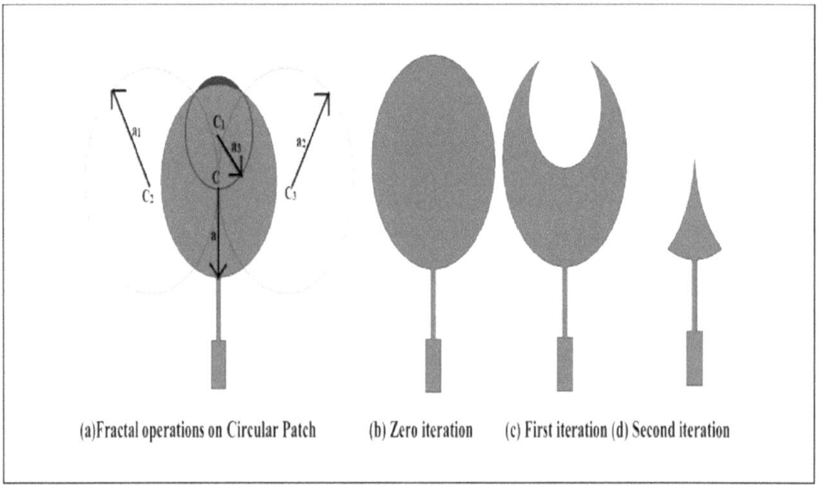

FIGURE 8.2 Antenna patch development using: (a) fractal operations on circular patch, (b) zero iteration, (c) first iteration, and (d) second iteration.

8.2.3 ANTENNA CONFIGURATION AND DESIGN DEVELOPMENT

The optimized parameters of the table lamp-shaped transformer fed, monopole fractal antenna with designations and its numerically measured values (in mm) are displayed in Table 8.1. Antenna front and rear view with designations are represented in the fourth column of the table in Figure 8.3(a) and (b).

TABLE 8.1 Optimized Design Dimensions of Table Lamp-shaped Antenna.

Parameters name	Designation	Value (mm)
Substrate length	L	26
Substrate width	W	13
50 Ω line length	L_1	3.0
50 Ω line width	W_1	6.0
Transformer fed length	L_2	7.78
Transformer fed width	W_2	0.9
Patch length	L_3	10.97
Radius of the circular Patch arc (table lamp base)	a	12
Radii of the two-Side arc (Table lamp two sides)	a_1=a_2	14
Radius of the top circular cut	a_3	7
Ground length	L_{gnd}	5.7
Ground width	W_{gnd}	13

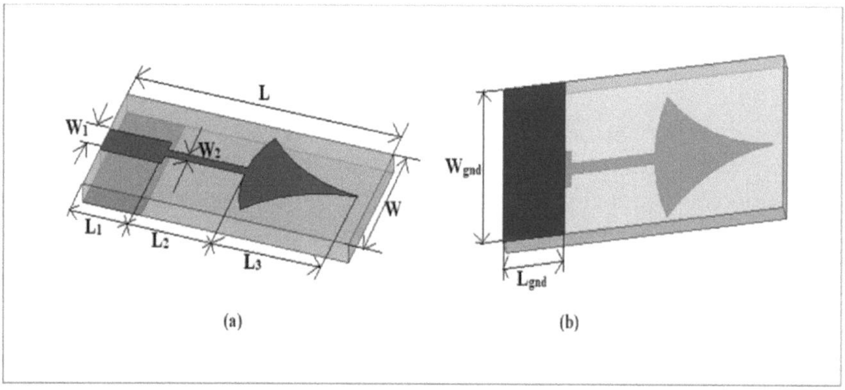

FIGURE 8.3 Antenna configuration with design parameters: (a) front view; (b) rear view.

8.2.4 MINIATURIZATION PROCESS

The initial size of the antenna is 48 × 48 mm (step 1) whereas the final size is 13 × 26 mm. A total 85.33% miniaturization of the fundamental size of an edge-fed circular microstrip antenna with complete ground is achieved by nine steps as represented in Figure 8.4 and their effects on reflection coefficients, antenna gain, radiation efficiency, and bandwidth are tabulated in Table 8.2.

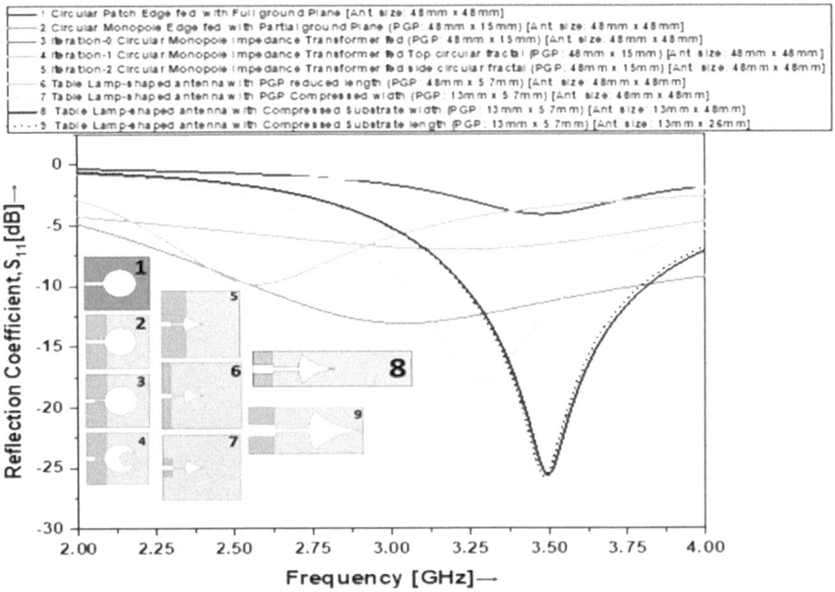

FIGURE 8.4 Miniaturization process and its effect on reflection coefficient plots.

It has a radiation efficiency of 50.54% without any resonance frequency and RF energy transmission/reception. Thus this antenna is converted into a monopole by reducing the ground length. Thus a resonance frequency at 3.04 GHz for a bandwidth 2.56 to 3.79 GHz with low reflection coefficient values below –10dB and has a good gain of 4.18 dBi for a PGP length of 15 mm. Initially, the ground plane and substrate size are equal to 48 × 48 mm. For lowering the reflection coefficient, S_{11} values below – 10 dB an impedance transformer is applied between the circular patch and 50 Ω feed line (step 3). Two levels of circular fractals are used for minimizing the size of the circular patch (steps 4 and 5). First fractal (iteration-1) cut is applied at the top with 7 mm radius and other two circular cuts of double radius, that is, 14 mm are cut along the two sides (step 5). The reflection coefficient and impedance matching are further improved by again decreasing the length of PGP to 5.7 mm (step 6) and compressing the width of PGP (CPGP step 7). Thus for fractal geometry with CPGP size 13 × 5.7 mm better impedance matching is achieved with reflection coefficient, S_{11} – 17.80 dB at resonant frequency 3.29 GHz within the bandwidth 3.04 to 3.59 GHz with maximum radiation efficiency 92.92% and good gain 2.69 dBi (step 7). Furthermore, the antenna size is reduced

TABLE 8.2 Step by Step Iniaturization Process.

Antenna patch Sub. W × L (mm^2)	Ground plane	Edge fed	Imp. transformer	BW (f_L-f_H^-) (GHz)	Res. Freq. f_r (GHz)	Peak gain G (dBi)	Ref. Coef. S_{11} (dB)	Rad. Eff. η (%)
	W × L (mm^2)	W × L (mm^2)	W × L (mm^2)					
Circular patch Full GP (48 × 48)	48 × 48	3 × 13.87	–	–	–	4.82	–	50.54
Circular monopole partial ground plane (PGP)(48 × 48)	48 × 15	3 × 13.87	–	2.56–3.79	3.04	4.18	−13.09	93.52
Iteration-0 circular Monopole transformer Fed (48 × 48)	48 × 15	3 × 6	0.9 × 7.78	–	–	3.95	–	91.30
Iteration-1 (Top cir. fractal) (48 × 48)	48 × 15	3 × 6	0.9 × 7.78	–	–	4.04	–	91.67
Iteration-2 (Side cir. fractal) (48 × 48)	48 × 15	3 × 6	0.9x7.78	–	–	1.78	–	73.89
Table lamp-shaped reduced PGP (48 × 48)	48 × 5.7	3 × 6	0.9 × 7.78	–	–	2.055	–	88.74
Table lamp-shaped CPGP (48 × 48)	13 × 5.7	3 × 6	0.9 × 7.78	3.04–3.59	3.29	2.69	−17.80	92.92
Table lamp-shaped compressed substrate width (13 × 48)	13 × 5.7	3 × 6	0.9 × 7.78	3.23–3.82	3.50	3.11	−25.63	94.77
Table lamp-shaped (13 × 26)	13 × 5.7	3 × 6	0.9 × 7.78	3.22–3.80	3.49	2.88	−25.80	95.90

by compressing the width of the substrate equal to CPGP, that is, 13mm, doing so results in improved S_{11} below –10 dB over the full 5G n78 band from 3.23 to 3.82 GHz and antenna gain becomes 3.11dBi (step 8). Finally, the size of the substrate is reduced along the length. The reduction of the substrate does not affect the antenna characteristics and parameters to a great extent (step 9). Thus proposed antenna was designed with a peak gain of 2.89 dBi, radiation efficiency of 95.90% at a resonant frequency of 3.49 GHz, and – 10 dB reflection coefficient, S_{11} for a bandwidth of 3.22–3.80 GHz.

8.3 RESULTS AND DISCUSSION

All the simulated resultant parameters of the table lamp-shaped monopole fractal antenna are displayed in Table 8.3. Antenna results in 85.33% miniaturized size, 2.88 dBi gain with radiation efficiency approximately 96% at 3.5 GHz. Low values of gain within the – 10 dB FBW (3.22–3.80 GHz) ensure that the antenna helps to keep the environment green and suitable candidate for 5G enabled smart cites n78 band mobile communications.

TABLE 8.3 Summary of Antenna Resultant Parameter.

Resultant parameters	Simulated antenna results	Desired parameters
Theoretical antenna size at f_0	48 × 48 mm	Evaluated as per design frequency, substrate permittivity, and thickness
Miniaturized antenna size	13 × 26 mm (0.15 λ0 × 0.30 λ0)	As minimum as possible
Miniaturization	85.33%	Not given
Design frequency,f_0	3.5 GHz	3.5 GHZ
Resonance frequency, f_r	3.485 GHz	Nearly equal to design frequency
Reflection coefficient at fr, S_{11}	−25.80 dB	<− 10 dB
−10dB bandwidth, BW	3.22–3.80 GHz	3.4–3.6GHz
Fractional bandwidth (FBW)	16.57%	As high as possible
Peak gain at f_0	2.88 dBi	Must be positive
Directivity, D	3.07 dBi	Must be positive
Radiation efficiency, η	95.90%	High and lies between
VSWR (<2)	3.21–3.82 GHz	<2
VSWR at f_r	1.10	Close to1

8.3.1 FABRICATED ANTENNA PROTOTYPE

The antenna is fabricated using film making, negative development, photolithography, and chemical etching process. Finally, a 50 Ω connector is soldered at the input excitation port on feed line carefully in the middle of the width without misalignments. The prototype images of the fabricated antenna along with its testing on Keysight Technologies Vector Network Analyzer (VNA), MODEL N9917A are displayed in Figure 8.5.

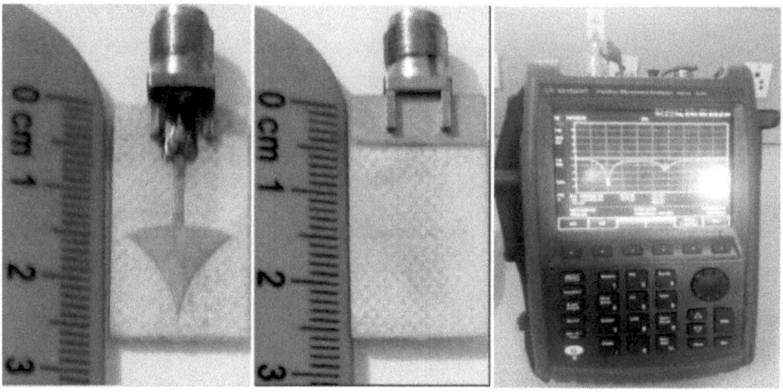

(a) (b) (c)

FIGURE 8.5 Prototype of table lamp-shaped antenna: (a) front image, (b) rear image, and (c) VNA measurements.

8.3.2 SIMULATED VS MEASURED REFLECTION COEFFICIENT, S_{11}, AND VSWR

The simulated results are found to be in close proximity to the measured results. A very slight frequency shift 0.05GHz has been observed in resonance frequencies of simulated and measured reflection coefficients, $S_{11,}$ and VSWR. The simulated and measured reflection coefficient curves and VSWR plots are represented in Figure 8.6(a) and (b), respectively.

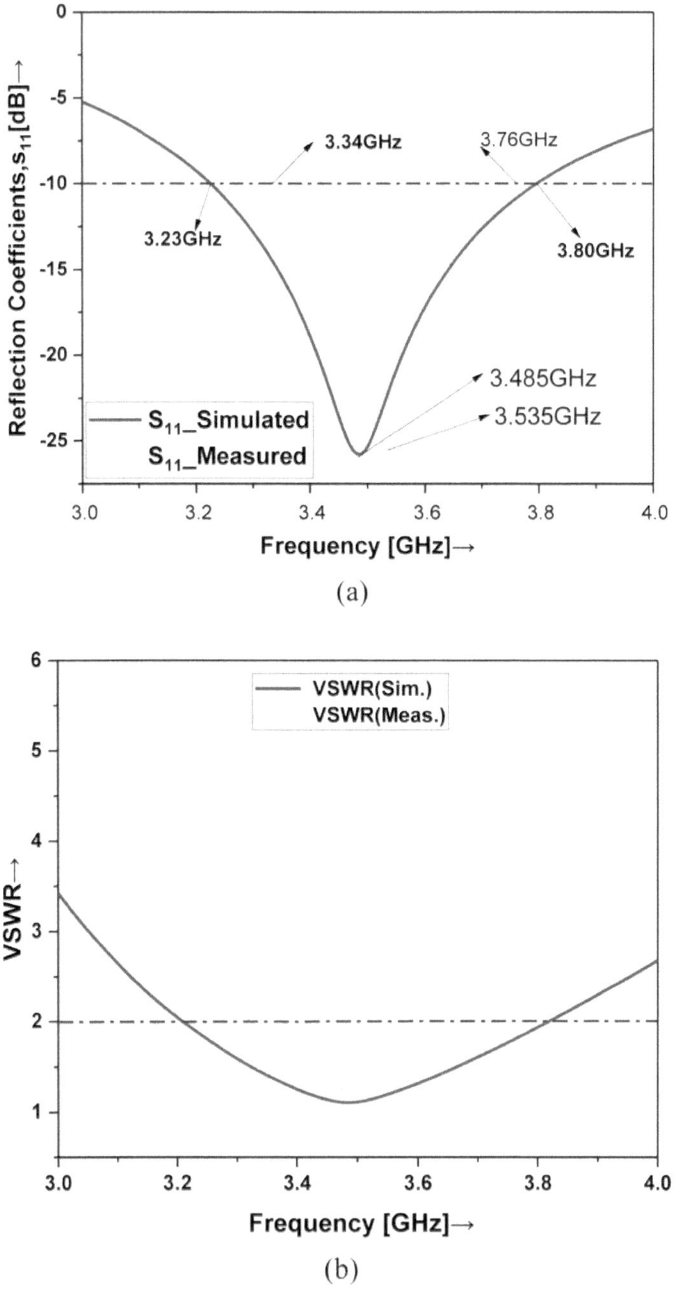

FIGURE 8.6 Measured vs. simulated: (a) reflection coefficients, S_{11} curves; (b) VSWR plots.

8.3.3 ANTENNA GAIN, DIRECTIVITY AND RADIATION EFFICIENCY

The antenna gain, directivity, and radiation efficiency of the table lamp-shaped antenna are plotted against frequency sweep 3 to 4GHz and these curves are plotted in the same graph as depicted in Figure 8.7. The antenna gain and directivity values lie between 2.5 and 3.5 dBi for N78 band 3.2 to 3.8 GHz band. The radiation efficiency of this antenna lies between 94 and –97%.

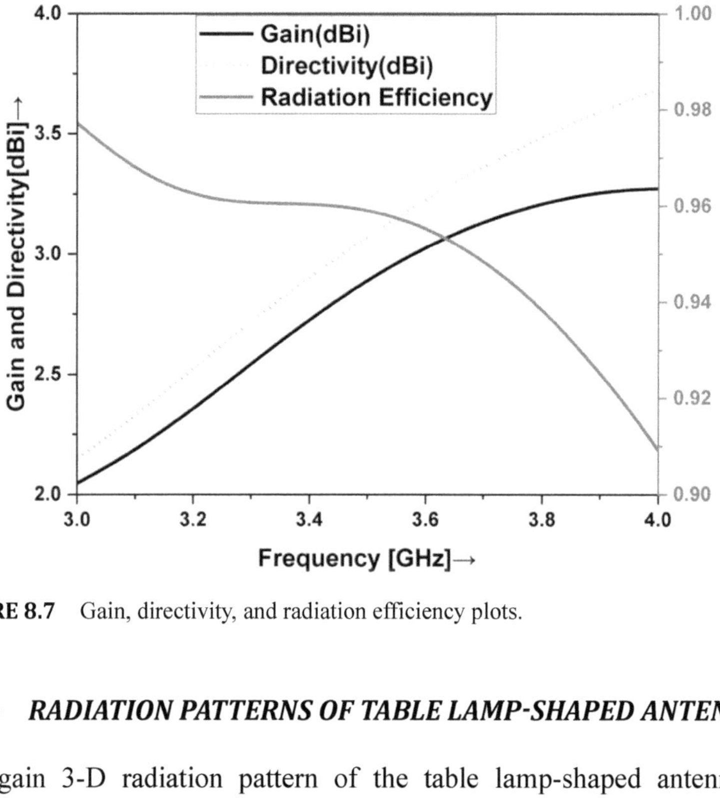

FIGURE 8.7 Gain, directivity, and radiation efficiency plots.

8.3.4 RADIATION PATTERNS OF TABLE LAMP-SHAPED ANTENNA

The gain 3-D radiation pattern of the table lamp-shaped antenna is omnidirectional and symmetrical around the theta (θ) and phi (Ø) plane as displayed in Figure 8.8(a) and in the E-plane the gain pattern is of the shape of alphabet 'O' while in the H-plane the gain pattern is of the shape of the numeral eight '8' as illustrated in Figure 8.8(b). The behavior of the antenna changes from unidirectional to bidirectional by reducing the area of the complete ground plane across the input excitation port.

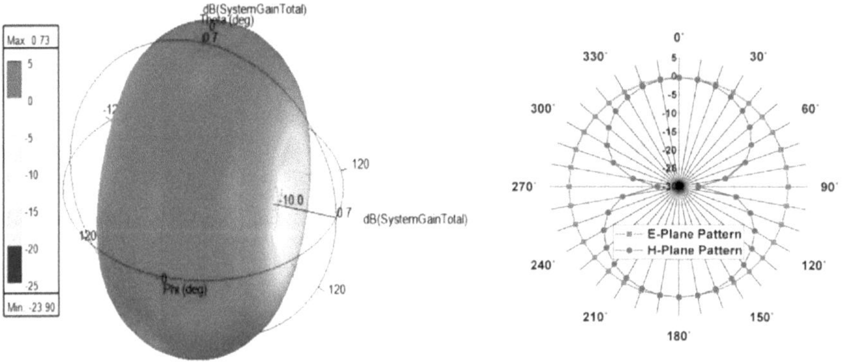

(a) (b)

FIGURE 8.8 Gain radiation patterns at frequency 3.5 GHz: (a) 3D-radiation pattern and (b) E-plane and H-plane polar plots.

8.3.5 NOVELTY OF PROPOSED ANTENNA

A comparative analysis of the proposed antenna in the present work with other relevant existing published works demonstrates superior performance of the proposed table lamp-shaped antenna, in terms of the number of operating bands, gain, reflection coefficient (S_{11}), antenna feeding techniques deployed[16] as well as size and bandwidth. Some important features of the antenna are also presented in Table 8.4.

8.4 CONCLUSION

A table lamp-shaped monopole two-iteration circular fractal antenna is fabricated and tested. Both measured and simulated results are found to be in excellent agreement. An 85.33% miniaturized antenna size with respect to basic design dimensions has been achieved. Low values of reflection coefficient, $S_{11,}$ and VSWR signify that a good quality of signal will be transmitted by the antenna. The antenna has good radiation efficiency, ≈ 96%, gaining 2.88 dBi within the bandwidth 3.22GHz–3.80GHz. Thus, the proposed antenna is suitable for WiMAX and 5G (n78 band) applications.

TABLE 8.4 Comparative Study of Proposed Work with Existing Literature Similar Antennas Designed at 3.5 GHz.

Ref	Size (W × L) (mm2)	Substrate	f_0 (GHz)	Feed	-10dB BW (MHz)	Peak Gain (dBi)	-S11 (dB)	VSWR	Features
[10]	32 46 × 1.6	FR-4	3.5	Edge	230/ 350/ 590 / 1210/ 1350	1.25/ 2.80/ 2.90 / 3.64/ 4.67	<-10dB	NG	-Monopole -12 PIN diodes -Frequency reconfigurable -5 no. of operating frequency bands -5 no. of beams -Several sub-6 GHz, GSM, 5G bands, UMTS Band, 4G-LTE Band WiMAX Bands, and WLAN Band
[9]	50 × 40 × 1.6	FR4	3.5	Proximity coupled	65.2	8.75	17.431	1.31	-Proximity Coupled -5G Applications—Microstrip with DGS -Unidirectional -Linearly Polarized
[11]	48 × 25.2 × 1.6	FR4	3.5	Offset Edge	410	7.16	30	NG	-Elliptical Patch -Monopole -5G and WiMAX Applications
[8]	100 × 100 × 1.6	FR4	3.5	Transformer	250	8.90	17	NG	- The patch antenna with side walls (right and left) and upper dielectric layer -WMAX Applications—Compact rectangular patch
[7]	63 × 51.2 × 4.5	FR4	3.5	Folded dipole	2330	6.20	30	1.10	-5G mobile communication -Sub-6GHz Applications
This work	13 × 26 × 1.6	FR4	3.5	Transformer	580	2.88	25.80	1.10	-Monopole -5G Applications (n78 band)

*NG, not given.

KEYWORDS

- **fractal**
- **lamp-shaped antenna**
- **M2M communication**
- **miniaturization**
- **n78 band**
- **transformer fed**

REFERENCES

1. Hassan, N. U.; Yuen, C.; Chen, X. Green Communications in Smart Cities. *Electronics* **2019,** *773* (8, 7), 1–3. DOI: https://doi.org/10.3390/electronics8070773
2. Guevara, L.; Cheein, F. A. The Role of 5G Technologies: Challenges in Smart Cities and Intelligent Transportation Systems. *Sustainability*, **2020,** *12* (6469), 1–15. DOI: https://doi.org/10.3390/su12166469
3. Nadeem, A.; Cho, H.-S.; Social-Aware Peer Selection for Device-to-Device Communications in Dense Small-Cell Networks. *Electronics* **2019,** *8* (6), 1–18. DOI: https://doi.org/10.3390/electronics8060670
4. Wei, Y.; Hwang, S. H. Spectrum Values in Suburban/Urban Environments Above 1.5 GHz. *Electronics* **2018,** *7* (12), 1–15. DOI: https://doi.org/10.3390/electronics7120401
5. Fatima, A. et al.; Virtual Machine Placement via Bin Packing in Cloud Data Centers. *Electronics* **2018,** *7* (12), 389, 1–22. DOI: https://doi.org/10.3390/electronics7120389
6. Lork, C. An Ontology-Based Framework for Building Energy Management with IoT. *Electronics* **2019,** *8* (5), 485, 1–15. DOI: https://doi.org/10.3390/electronics8050485
7. Zahid, M. et al.; Electricity Price and Load Forecasting using Enhanced Convolutional Neural Network and Enhanced Support Vector Regression in Smart Grids. *Electronics* **2019,** *8* (2), 122, 1–32. https://doi.org/ 10.3390/electronics8020122
8. Cirik, F.; Yildirim, B. S. Analysis and Design of a 3.5-GHz Patch Antenna for WiMAX Applications. In: *International Journal of Microwave and Wireless Technologies*; Cambridge University Press and the European Microwave Association, **2014**; pp 1–8. DOI: https://doi.org/10.1017/S1759078714001238
9. Paragya, D.; Siswono, H. 3.5 GHz Rectangular Patch Microstrip Antenna with Defected Ground Structure for 5G. *ELKOMIKA: Jurnal Teknik Energi Elektrik, Teknik Telekomunikasi, & Teknik Elektronika 8* (1), 31–42. DOI : http://dx.doi.org/10.26760/elkomika.v8i1.31
10. Ahmad, I. et al. A Pentaband Compound Reconfigurable Antenna for 5G and Multi-Standard Sub-6GHz Wireless Applications. *Electronics* **2021,** *10* (2526), 1–17. https://doi.org/10.3390/electronics10202526

11. Ferdous, N. et al. Design of a Small Patch Antenna at 3.5 GHZ for 5G Application. *Conference Series: Earth and Environmental Science* **2019,** *268*, 012152, 1–4. DOI:10.1088/1755–1315/268/1/012152
12. An, W. et al. Low-profile and Wideband Microstrip Antenna with Stable Gain for 5G Wireless Applications. *IEEE Antennas Wireless Propagation Lett.*, **2018,** *17* (4), 621–624. DOI: 10.1109/lawp.2018.2806369.
13. Khandelwal, M. K. Defected Ground Structure: Fundamentals, Analysis, and Applications in Modern Wireless Trends. *Int. J. Antennas Propagation* **2017,** *1*, 1–22. DOI: https://doi.org/10.1155/2017/2018527
14. Balani, C. A. *Antenna Theory*, 4th edn.; Jhon Wiley and Sons, 2009.
15. BernaTableLamp-UrbanLadder. https://www.urbanladder.com/products/berna-table-lamp
16. Bisht, S.; Saini, Prakash, D. V.; Nautiyal B.; Study The Various Feeding Techniques of Microstrip Antenna Using Design and Simulation Using CST Microwave Studio. *Int. J. Emerg. Technol. Adv. Eng.*, **2014,** *4*, (9), 318–324.
17. Pradhan, D.; Tun, H. M. Circular-MSPA: Design and Analysis of Applications Intended for 5G Environment. *J. Netw. Secur. Comput. Netw.* **2023**, *9* (1), 14–19). https://doi.org/10.46610/jonscn.2023.v09i01.002
18. Oo, W. M.; Tun, H. M.; Nway, T. M.; Pradhan, D.; Sahu, P. K.; Naing, Z. M. Design, Analysis and Fabrication of Dual Band Microstrip Patch Antenna for (L2) Band GPS and WiFi Applications. In *2022 International Conference for Advancement in Technology (ICONAT)*; IEEE. **2022**; pp 1–5.
19. Tun, H. M.; Lin, Z. T. T.; Pradhan, D.; Sahu, P. K. Slotted Design of Rectangular Single/Dual Feed Planar Microstrip Patch Antenna for SISO and MIMO System. In: *2021 International Conference on Electrical, Computer and Energy Technologies (ICECET)*; IEEE, **2021**; pp 1–6.
20. Pardhan, D. (2019). Design of Extended Circular Patch with Rectangular Stub and Circular Slit Used For Ultra Wide Band Application (X-Band). *IOSR J. Appl. Phys. (IOSRJAP)* **2019,** *11* (4), 14–24.
21. Pardhan, D. Circular Patch With Circular Slit Patch Antenna Used for Ultra Wide Band Application. *Int. J. Electr. Electron. Data Commun.* **2017,** *5* (2), 84–87.
22. Pradhan, D.; Tun, H. M. Security Challenges: M2M Communication in IoT. *J. Electr. Eng. Autom.* **2022,** *4* (3), 187–199.

CHAPTER 9

A Tri-band Meander Line Fed Alpha-Numeric Antenna for Military Band, WiFi, and 5G Wireless Green Communications

ATUL VARSHNEY[1], VIPUL SHARMA[1], T. MARY NEEBHA[2], and N. PRASANTHI KUMARI[3]

[1]*ECE Department, FET, Gurukula Kangri (Deemed to be University), Haridwar, Uttarakhand, India*

[2]*ECE Department, Karunya Institute of Technology and Science, Coimbatore, Tamilnadu, India*

[3]*Electrical Department, SoE, UPES, Dehradun, Uttarakhand, India*

ABSTRACT

In this chapter a monopole, miniaturized, tri-wideband, alpha-numeric microstrip antenna for military band, WiMAX, and 5G wireless green communications for smart cities has been proposed, fabricated, tested, and investigated. An alpha-numeric structure has been used to design the antenna. The design comprises one alphabet 'V' and a numeral '5.' The antenna utilizes a PGP with reduced length to give the antenna wideband performance. Conventionally used, coaxial feed, inset feed and edge feed techniques are most common and outdated from narrow bands. Nowadays, a meander line is alternatively used for antenna feeding. Meander line results in an excellent impedance matching at designed frequency and

5G Green Communication Networks for Smart Cities. Devasis Pradhan, Rajeswari, Hla Myo Tun, Naw Khu Say Wah, & Thandar Oo (Eds.)

consequently, the proposed antenna resonates precisely at the operating frequency with −39.75 dB reflection coefficient value. The antenna has less than − 10 dB reflection coefficient (S11) in three frequency wide bands, that is, (<2.0–3.64 GHz), (4.52–5.36 GHz) and (6.2–6.7 GHz). Therefore, the antenna is suitable for ISM band applications like Wi-Fi 4 (IEEE802.11n standard), Wi-Fi 5 (IEEE802.11ac standard) and Wi-Fi 6 (IEEE 802.11ax standard), WLAN, WiMAX (2.3–2.4/2.5–2.6/3.4–3.6 GHz) and wireless applications like WCDMA (2.1 GHz), 3G (2.1 GHz) and 4G LTE (2.1/2.3/2.5/2.7 GHz), 5G frequency bands n1 (2.1 GHz), n2 (1.9 GHz), n3 (1800 GHz), n7 (2.6 GHz), n38 (2.57–2.62 GHz), n41 (2.94–2.69 GHz), and n78 (3.2–3.67 GHz). The work reports high radiation efficiency 93.81% and 3.59 dBi improved gain even with a total size reduction of 30.3% as compared to that of the fundamental rectangular patch antenna designed at frequency 3.4 GHz. The Electric LC equivalent circuit has been generated for the proposed antenna and the measured radiation patterns in E and H-planes for 3.40, 5.62, and 6.88 GHz have been displayed along with simulated patterns and they are found in very good agreements.

9.1 INTRODUCTION

The concept of microstrip antenna was first introduced in 1950 and then first came in the printed circuit board form in 1970. After that antenna became very popular.[1] The microstrip antennas are the most popular one because of their small size, low weight and cost, low volume, and portability and a single antenna can serve for single to multiple applications. Nowadays, the antennas are the backbone of every wireless communication system to transmit and receive the electromagnetic signals. For this purpose, many antennas have been designed and developed. In the present scenario as per the government policies, many devices and systems are deployed to make many 5 G-enabled smart cities for green communications. Millions of connections are required for trillions of message transfers among the wireless systems and devices for machine-to-machine (M2M) and device-to-device (D2D) communications to felicitate the smart city services. For wireless communications, rectangular, circular, elliptical, square, and triangular microstrip antennas are the roots of building any advanced fractal microstrip antenna. A series of antennas are available still the research is going on to reduce the overall area or miniaturize the size of antenna with enhanced gain, wide band, multiband, and ultra-wide bandwidth. With the modernization in microwave technology development, more requirements for miniaturized microstrip antennas must be needed to design. For MMIC/MIC low radiating patch area is also required with the possibility to add more numbers of passive components on its remaining surface area. To comply with this a miniaturized alpha-numeric antenna (ANA) has been designed and investigated in this article.

The narrow band antenna has bandwidth <500 MHz, the wideband antenna has a bandwidth > 500 MHz but <7.5 GHz whereas the ultra-wideband (UWB) antenna is defined as one that has−10 dB fractional bandwidth (FBW) of 7.5 GHz according to FCC.[2] The bandwidth allocated to UWB is from 3.1 GHz to 10.6 GHz by the Federal Communications Commission (FCC) in 2002.[3] Sharma and Sandhu have designed a PGP-based, edge fed, monopole, wideband with V-shaped slots, omnidirectional rectangular slotted patch antenna for wireless applications. The V-slotted antenna exhibits 3.87 dB and 4.67 dB gains at 2.45 GHz and 7.52 GHz frequencies respectively.[4] Sharma has presented two trapezoidal patch antennas, first with a V-shaped slot and second with an inverted V-shape slot. The first antenna exhibits an impedance bandwidth in the frequency range of 5.1–5.94 GHz, that is, 15.6% of the center frequency for WLAN/WiMAX applications, and the second antenna exhibits an impedance bandwidth of 30% in the frequency range from 3.54 to 4.85 GHz for radar and Japan UWB application.[5] Ram Krishna and Kumar have designed and investigated a microstrip fed open V-shaped slot antenna for wideband dual slant polarization. The antenna (76.25 × 52.25 mm) is dual fed, exhibits bandwidth 2.25–12 GHz, and gain lies between 3 and 5 dB with more than 20 dB isolation.[6] Osklang et al. have proposed a miniaturized size (17 × 23.5 mm) triband compact printed antenna for WLAN and WiMAX applications.[7] The meander line idea is to fold the conductors back and forth to make the overall antenna shorter.[7,8] A meander line antenna can be realized by bending the conventional linear monopole antenna to decrease the size of the antenna.[9] The meander line length and width and shape will affect the performance parameters of an antenna.[10] Wang et al. have designed a miniature antenna by etching a '15' I-shaped resonating- ring from the main patch by condition of miniaturization to maintain the bandwidth and the maximum gain. The overall area of the designed antenna at 5.8 GHz is decreased by 35.2% compared with the conventional rectangular microstrip antenna.[11] Bird has detected the misuse of Return Loss in the literature and gives the correct definition of return loss and input reflection coefficient (S11).[12] A. Varshney et al. have designed a monopole antenna with a tri-blade arm microstrip antenna loaded with split ring resonator (SRR) triplet. The antenna exhibits high gain and ultra-wideband behavior.[13,14]

In the proposed work a monopole, miniaturized, hybrid (edge plus meander line) fed, ANA is investigated, fabricated, and tested for WiMAX, Wi-Fi, and Wireless applications. The antenna utilizes a defected ground structure that plays a major role to miniaturize the original ANA to 30.3% with improved gain and reflection coefficient values. Paragraph: use this for the first paragraph in a section, or to continue after an extract.[27–30]

9.2 EXPERIMENTAL METHODS AND MATERIALS

The ANA is designed at 3.4 GHz frequency and devised on a piece 28.45 × 33.4 mm of FR-4 substrate with loss tangent 0.02. The said substrate has a permittivity 4.4 and a thickness of 1.6 mm.

9.2.1 DESIGN EQUATIONS

The ANA is developed from the most fundamental rectangular microstrip patch antenna. The design analysis was carried out by using the following standard design equations[4,15,16,17];

i) Width of the radiating Patch

$$W = \frac{c}{2f_r}\sqrt{\frac{2}{\varepsilon_r + 1}} \tag{9.1}$$

Where,

c: velocity of light, 3 × 108 m/s,

εr: dielectric constant of the substrate.

fr: resonant frequency of antenna

ii) Effective Dielectric constant of the rectangular patch

$$\varepsilon_{reff.} = \frac{\varepsilon_r + 1}{2} + \frac{\varepsilon_r - 1}{2}\left(1 + \frac{12h}{W}\right)^{-0.5} \tag{9.2}$$

iii) The effective length of rectangular patch at the resonance frequency

$$L_{eff} = \frac{c}{2\varepsilon_r\sqrt{\varepsilon_{reff.}}} \tag{9.3}$$

iv) The extension length of rectangular patch

$$\Delta L = 0.412h\frac{(\varepsilon_{reff.} + 0.3))(\frac{W}{h} + 0.264)}{(\varepsilon_{reff.} - 0.258)(\frac{W}{h} + 0.8)} \tag{9.4}$$

v) The length "L" of rectangular patch

$$L = L_{eff^{.}} - 2\Delta L \tag{9.5}$$

vi) Microstrip Feed Length

$$L_f = \frac{\lambda_g}{4} \tag{9.6}$$

Whzere,

$$\lambda_g = \frac{c}{f_r\sqrt{\varepsilon_{reff.}}} \tag{9.7}$$

vii) Substrate width

$$W_{sub^{.}} = W + 6h \tag{9.8}$$

viii) Substrate Length

$$L_{sub^{.}} = L + L_f + 3h \tag{9.9}$$

ix) Feed line width

$$W_f = \frac{7.48h}{e^{(Z_0\frac{\sqrt{\varepsilon_r + 1.41}}{87})}} - 1.25t \tag{9.10}$$

Where,
Z_0=Characteristics impedance of feed line
t = thickness of copper conductor

9.2.2 DESIGN FLOW AND DEVELOPMENT

All parameters of the rectangular patch antenna using equation (9.1) through equation (9.10) were calculated. Then, antenna geometry designed in HFSS software using evaluated parameters. Afterward, required modifications according to the desired ANA designs have been incorporated as mentioned in Figure 9.1.

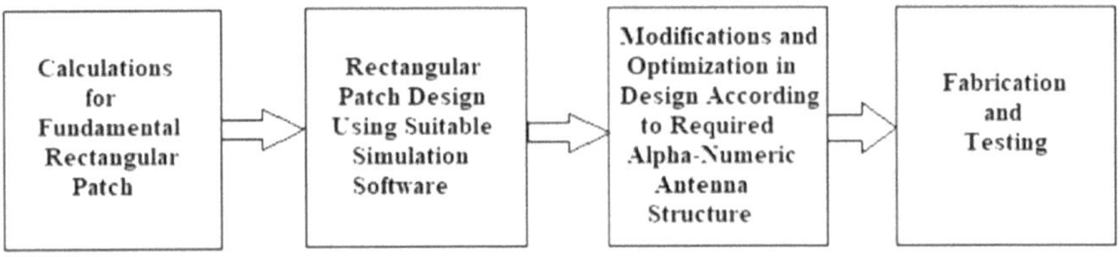

FIGURE 9.1 Block diagram of design flow and development.

9.2.3 DESIGN OF ANA

An ANA is a combination of an English alphabet 'V' and a numeric number '5'. Thus, all necessarily required changes have been made and optimized using HFSS software tool properties, on basic rectangular patch antennas to miniaturize the antenna into ANA shaped. The resultant modified antenna geometries are shown in Figure 9.2. An ANA utilized a PGP to make the antenna performance wideband and a meandered line for excellent impedance matching. The reduced length ground near the feed (arrow direction mentioned in Figure 9.2(b)) changes the antenna geometry into monopole antenna. The dimensions of the simple ANA of the designed antenna are 36.45 × 37.40 mm. An ANA resonates exactly at the designed frequency, that is, at 3.4 GHz with − 17.69 dB reflection coefficient over the range 2.87 to 3. 91 GHz. The achieved gain of simple ANA is 3.30 dBi with a radiation efficiency of 91.35%.

9.2.4 DESIGN OF MINIATURIZED ANA

An ANA antenna has low gain and low reflection coefficient value lower than −10 dB with a larger occupied area 36.45 × 37.40 mm. These parameters are improved to a great extent by reducing the overall antenna size. Hence, a miniaturized version of the ANA has been obtained. The optimized dimensions of the miniaturized ANA are 28.45 × 33.4 mm. This miniaturization results in a total 30.3% area reduction in the initially designed ANA structure without change in the original design structure as illustrated in Figure 9.3.

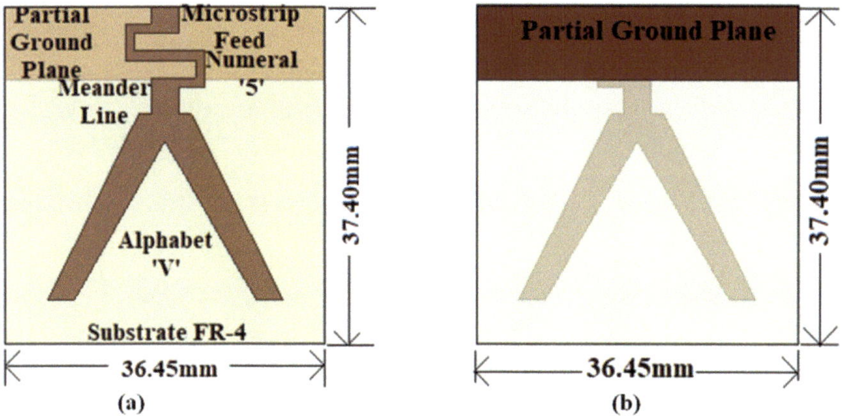

FIGURE 9.2 Alpha-numeric antenna (ANA): (a) radiating patch and (b) partial ground plane.

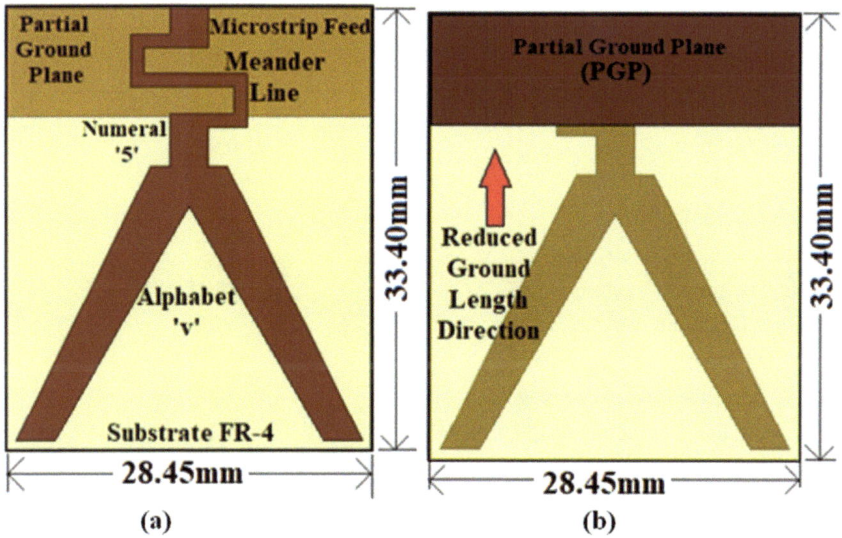

FIGURE 9.3 Miniaturized ANA: (a) radiating patch and (b) reduced ground structure.

9.2.5 OPTIMIZED DIMENSIONS OF MINIATURIZED ANA

The three-dimensional trimetric view and symbolic dimensional views have been represented in Figure 9.4(a–b) and all symbols with their designation and optimized parameter values have been tabulated in Table 9.1.

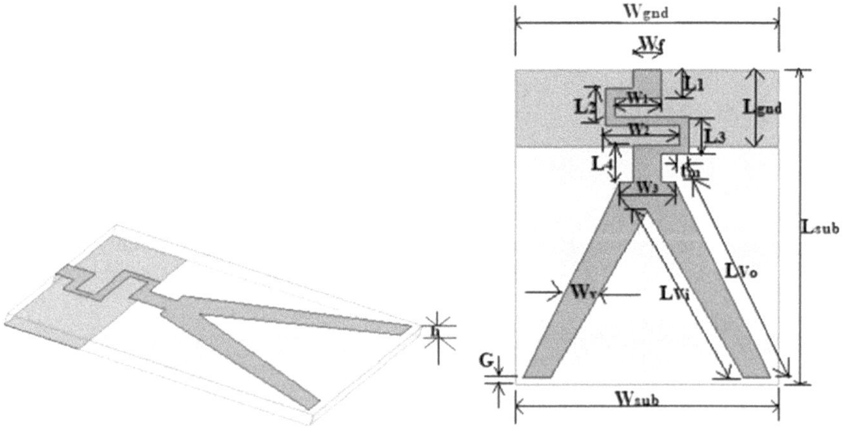

(a) (b)

FIGURE 9.4 Dimensional view of miniaturized ANA: (a) 3-D trimetric view and (b) parametric dimensional view.

TABLE 9.1 Optimized Parameter Table.

Designation/ Description	Value (in mm)
Metal thickness, t	0.035
Substrate thickness, h	1.6
Substrate width, Wsub	28.45
Substrate length, Lsub	33.4
Width of ground structure, Wgnd	28.45
Length of ground structure, Lgnd	8.2
50 Ω Microstrip feed width, Wf	3.0
50 Ω Feed line first section length, L_1	3.0
Meander line first section length, L_2	4.0
Meander line second section length, L_3	4.0
50 Ω Microstrip feed second section length, L_4	4.0
Meander line first section width, W_1	5.0
Meander line second section width, W_2	8.0
Width of top head of letter 'V', W_3	6.0
Meander line thickness, tm	1.0
Inner length of letter 'V', l_{vi}	20.456
Outer length of letter 'V', L_{V0}	23.088
Width of letter 'V', W_V	3.0
Gap between letter 'V' arm and substrate, G	0.8

9.2.6 ADSOR PROTOTYPE OF MINIATURIZED ANA

Miniaturized ANA is finally fabricated with optimized parametric values mentioned in Table 9.1 by film making, negative preparation, photolithography and chemical etching process. The prototype of the proposed antenna has been illustrated in Figure 9.5(a). Miniaturized ANA is finally fabricated with optimized parametric values mentioned in Table 9.1 by photolithography and chemical etching process. The prototype of the proposed antenna has been illustrated in Figure 9.5(a). The photographs of laboratory tested parameters and measurements of reflection coefficient, Gain, E-plane and H-plane radiation patterns are represented in Figure 9.5(b).

FIGURE 9.5 Prototype of miniaturized ANA: (a) front and rear view; (b) laboratory testing photographs of ANA.

9.2.7 RLC ELECTRICAL EQUIVALENT OF ANA

There are many methods to develop an electrical equivalent circuit of the microstrip patch antenna out of which the equivalent circuit of proposed miniaturized ANA is generated using an electric inductive capacitive (ELC) technique.[13,18,19] The proposed ANA is modeled from the simulated reflection coefficient (S_{11}) value. First the S-parameter is converted into Y-parameter.[20] Then the electrical equivalent values of passive components at resonance are reproduced by electrical theory of parallel RLC resonance circuits.[21] The final generated electrical equivalent circuit is illustrated in Figure 9.6.

$$Y_{11} = G + j\,B \tag{9.11}$$

$$G = \frac{1}{R} \tag{9.12}$$

$$B = (2\pi frC - \frac{1}{2\pi f_r L} \tag{9.13}$$

$$Q = \frac{f_r}{BW} = 2\pi RC \tag{9.14}$$

$$-10\text{ dB fractional Bandwidth,}\quad FBW = \frac{f_H - f_L}{f_c} \tag{9.15}$$

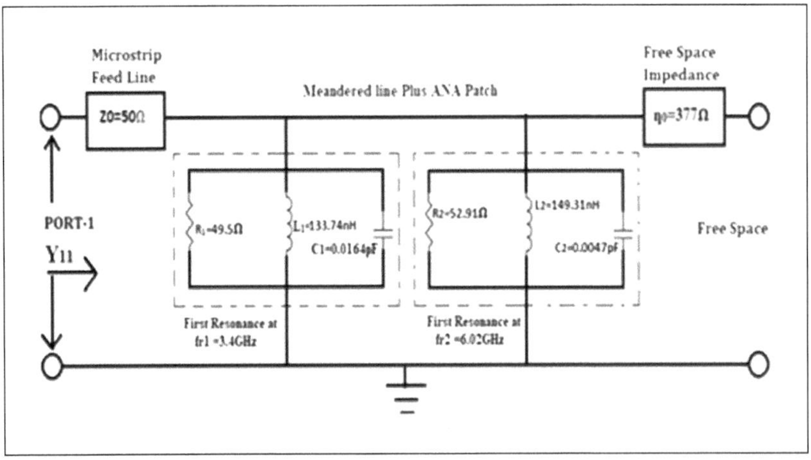

FIGURE 9.6 Equivalent circuit of miniaturized ANA: Electric LC (ELC).

9.3 RESULTS AND DISCUSSION

9.3.1 MINIATURIZATION PROCESS AND ITS EFFECTS ON PRIMARY ANA

The miniaturization process and its effect on reflection coefficient plots and other antenna parameters have been compared and analyzed. Initially, primary ANA is designed with complete ground and then the ground plane length is reduced near the RF excitation feed line. This will enhance the reflection coefficient below − 10 dB and the nature of the S_{11} plot becomes a dual wideband nature. Further, the width of the PGP is compressed, resulting in a compressed partial ground plane (CPGP). This will add one more band to the reflection coefficient curve. Finally, the substrate length and width are also compressed. This will result in improved impedance matching in miniaturized ANA as shown in Figure 9.7. Hence, miniaturized ANA results in about −40 dB in reflection coefficient[12] value at resonance frequency 3.4 GHz over the same – 10 dB fractional bandwidth range 2.87 to 3.91 GHz in both the cases. Table 9.2 compares the simulated parameters of primary ANA and Miniaturized ANA. Comment columns in the table explain about the resultant parameter enhancement after the 30.3% reduction in area.

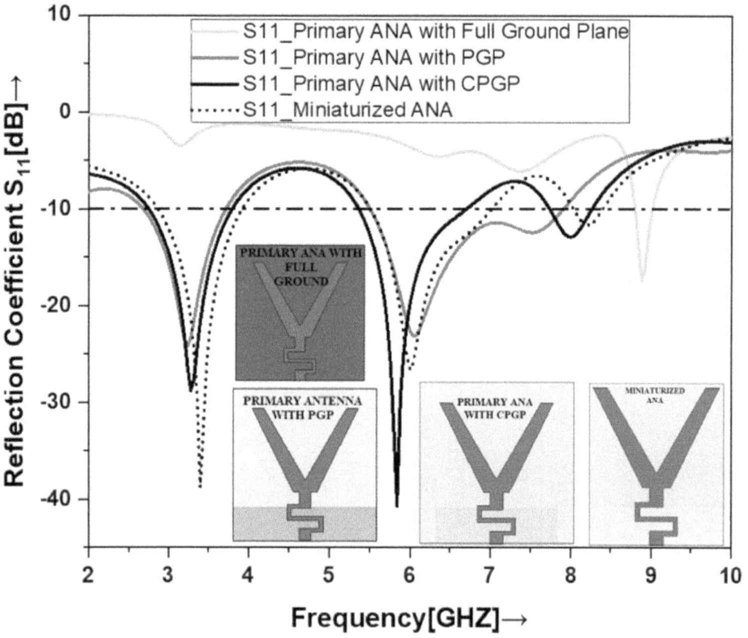

FIGURE 9.7 Reflection coefficients curves of initial ANA vs. miniaturized ANA.

9.3.2 REFLECTION COEFFICIENTS CURVES OF SIMULATED VS. MEASURED ANA

A comparative view of simulated and measured reflection coefficient curves is shown in Figure 9.8. Table 9.3 illustrates the resultant parameters of simulated and measured miniaturized ANA. The measured values of reflection coefficient curves are found in close proximity to the simulated S_{11} plot. The first band of the measured reflection coefficient is found

TABLE 9.2 Performance Comparisons of Simple ANA and Miniaturized ANA.

Antenna type	Size (mm^2)	Bands (− 10 dB FBW)	Resonance frequencies	S_{11} [dB]	Gain (dBi)	Rad. Eff. (%)
1. Primary ANA With Full Ground	36.45 × 37.40	1(8.80–9.01 GHz)	8.87 GHz	− 17.69	–	–
2. Primary ANA With PGP	36.45 × 37.40	2(2.87–3.91 GHz), & (5.04–7.39 GHz)	3.40 GHz, 5.54 GHz	− 17.69, − 33.42	3.22	91.25%
3. Primary ANA With CPGP	36.45 × 37.40	3(2.75–3.79 GHz), (5.36–6.70 GHz) & (7.73–8.26 GHz)	3.28 GHz, 5.84 GHz' 8.0 GHz	− 28.60, − 40.50, − 12.89	3.43	93.05%
4. Miniaturized ANA	28.45 × 33.40	3(2.86–3.92 GHZ), (5.50–7.02 GHz), & (8.03–8.40 GHz)	3.40 GHz, 6.02 GHz, 8.21 GHz	− 39.78, − 25.50, − 11.74	3.59	93.81%
Comment on Antenna2 and Antenna 4	Reduced by 30.3%	One band increased	Second resonance frequency shifts toward right	Impedance matching improves at designed/First resonance frequency	Improved	Radiation Efficiency Increased

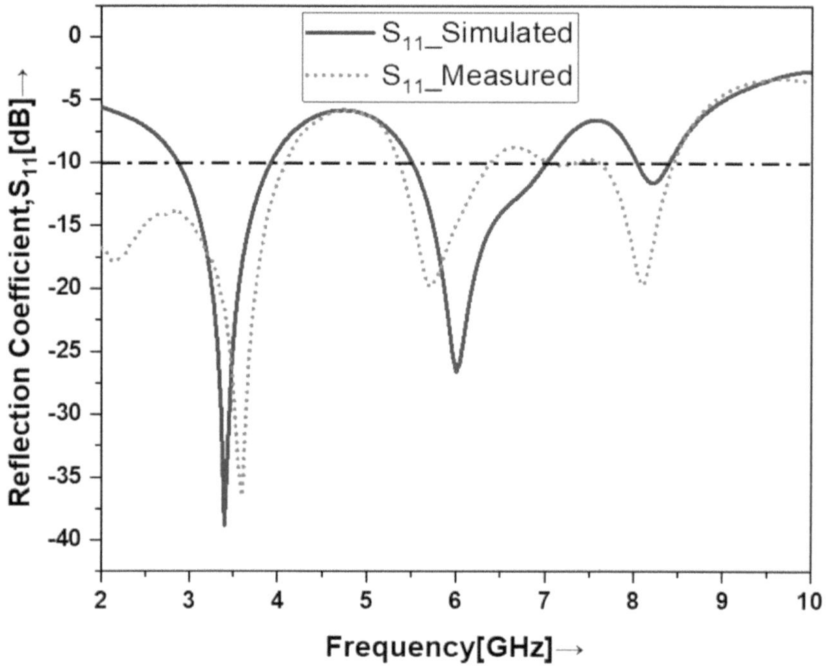

FIGURE 9.8 Reflection coefficient curves of simulated vs. measured ANA.

TABLE 9.3 Performance Comparisons of Simulated and Measured ANA.

Resultant Parameter	Miniaturized ANA (Simulated)	Miniaturized ANA (Measured)	Comment
No. of bands	3	3	Unchanged
Resonance frequencies	3.40 GHz, 6.02 GHz, 8.21 GHz	3.60 GHz, 5.72 GHz, 8.02 GHz	First main resonating frequency shift toward right, second shifts left and third shifted toward left hand sides
S_{11}[dB]	− 39.78, − 25.5, − 11.74	− 36.54, − 19.82, − 19.60	Decreased at all frequencies by acceptable range
− 10 dB FBW	2.87–3.92 GHz, 5.49–7.03 GHz, 8.03–8.40 GHz	<2–3.67 GHz, 4.52–5.36 GHz, 6.2–6.7 GHz	FBW Improved for first resonance and an extra band noticed that is very efficient for various wireless applications
Gain	3.59 dBi	3.60 dBi	Almost same

wider than the simulated reflection coefficient of the first band. These deviations in the plots are due to the fabrication process error, soldering of the connector and substrate imperfection of dimensions. The second band is found little left sifted while the third band is found with improved wide reflection coefficient values below − 10 dB.

9.3.3 RADIATION PATTERN

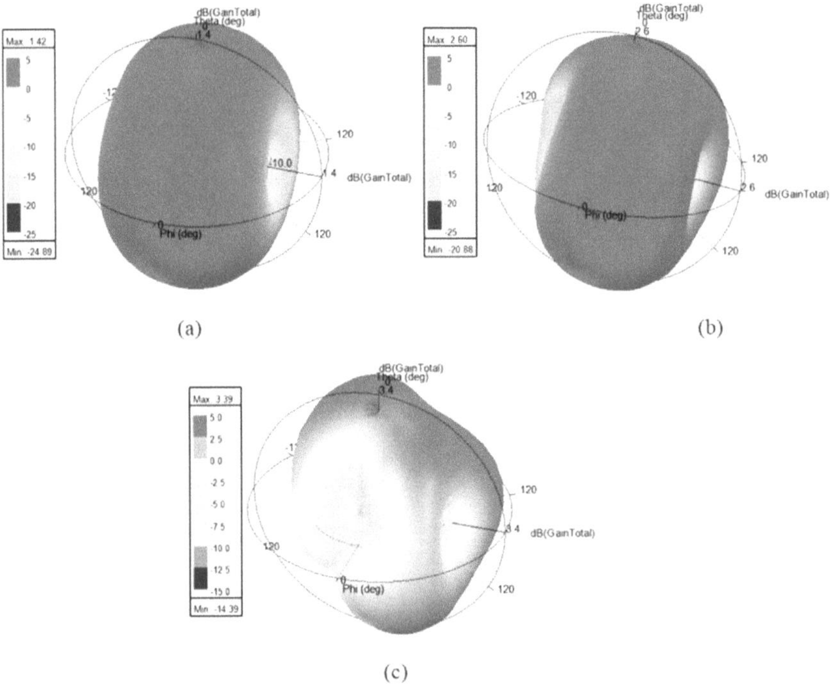

FIGURE 9.9 Radiation gain patterns of ANA at: (a) 3.40 GHz, (b)5.62 GHz, and (c) 6.88 GHz.

The radiation pattern of the ANA is omnidirectional in the first band (2.86 –3.92 GHz) as illustrated in Figure 9.9(a) at 3.40 GHz. The ANA has a maximum peak gain of 1.42 dB or 3.57 dBi at a resonant frequency of 3.40 GHz. At rest, two bands (5.49–7.03 GHz and 8.03–8.40 GHz) radiation pattern disturbs its omnidirectional shape, and improved gains 4.75 and 5.54 dBi are obtained at the other two measured frequencies 5.62 and 6.88 GHz, respectively. The E-plane and H-plane patterns of ANA are plotted on the same polar plot at frequencies 3.40, 5.62, and 6.88 GHz as shown in Figure 9.10(a–f).

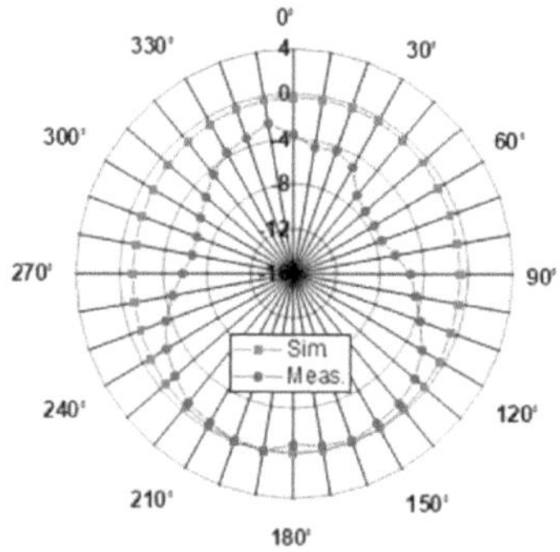

(a) E-Plane Pattern @3.40 GHz

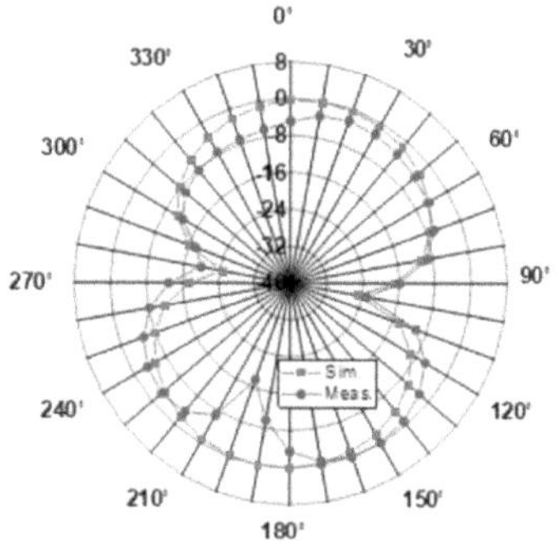

(b) H-Plane Pattern @3.40 GHz

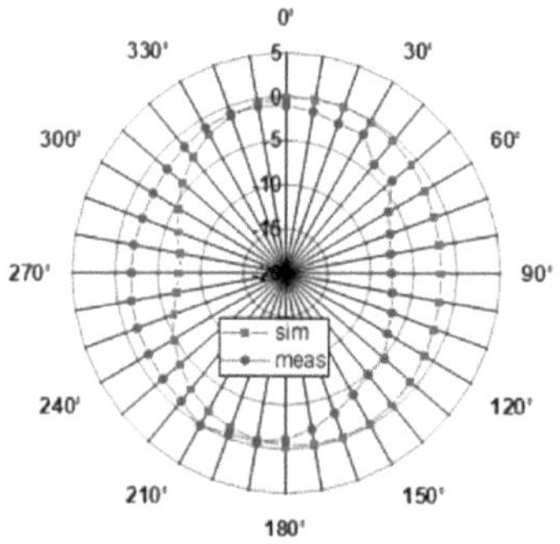

(c) E-Plane Pattern @5.62 GHz

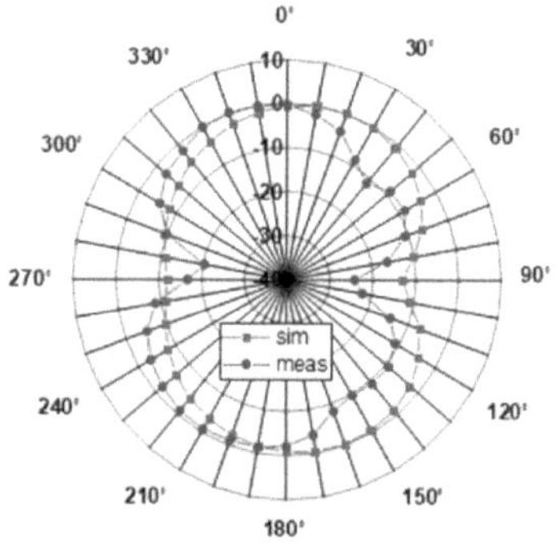

(d) H-Plane Pattern @5.62 GHz

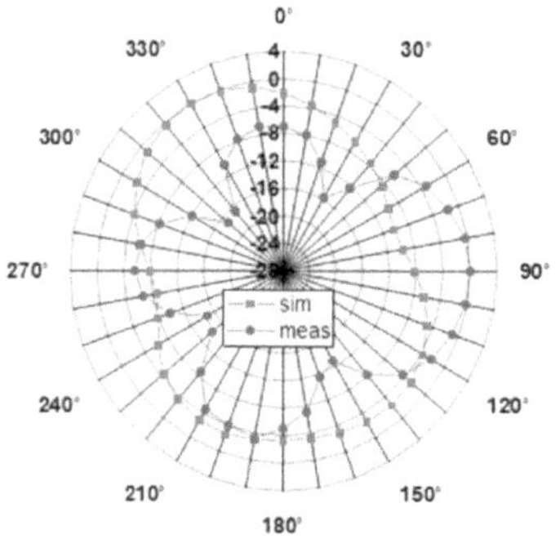

(e) E-Plane Pattern @6.88 GHz

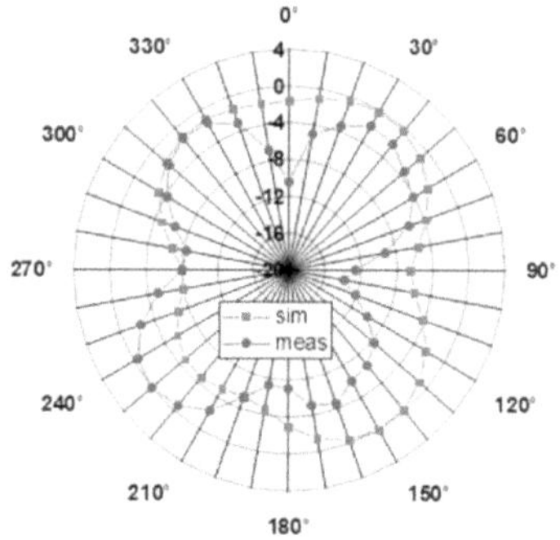

(f) H-Plane Pattern @6.88 GHz

FIGURE 9.10 E-plane (φ=0°) and H-plane (φ=90°) radiation patterns of ANA at 3.40 GHz.

9.3.4 ANTENNA GAIN, DIRECTIVITY AND RADIATION EFFICIENCY

The antenna parameters (gain, directivity, and radiation efficiency) variations with frequency are depicted in Figure 9.11. The highest value of antenna gain 6.47 dBi found at frequency 8.04 GHz. Antenna radiation efficiency lies between 78 and 97%.

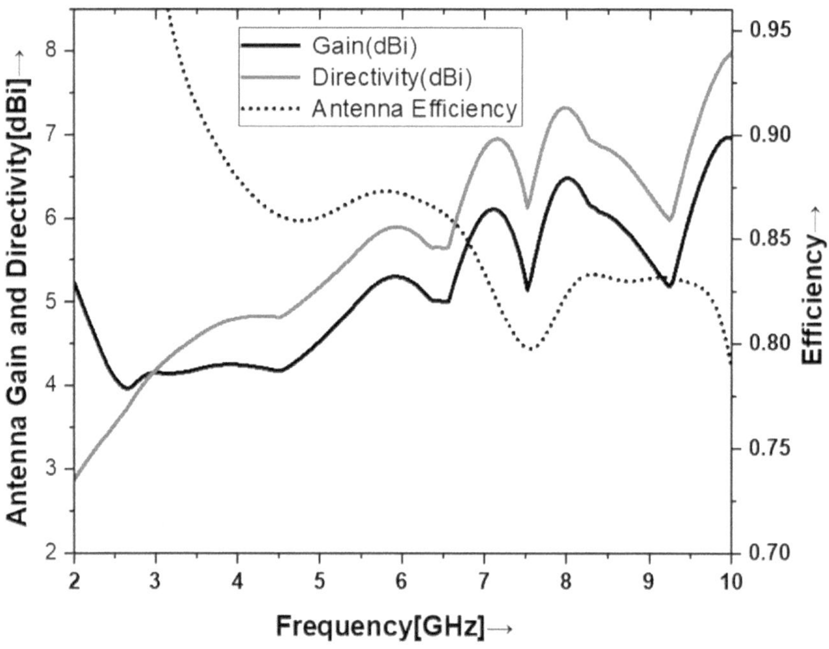

FIGURE 9.11 Antenna parameters plots.

9.3.5 CONTRAST OF PROPOSED WORK WITH RECENTLY PUBLISHED WORK

The proposed ANA has a miniaturized size that comprises of hybrid feeding methodology (edge plus meander line) with high gain and wide-band performance for starting two bands and narrow band performance for the third band. Table 9.4 illustrates the performance of the proposed ANA with the recently published antennas in near domains.

TABLE 9.4 Performance Comparisons Proposed ANA with Recently Published Antennas.

Ref.	f_r(GHz)	Freq. band	Size	FBW (%)	Gain (dBi)	Features
[17]	3.30	1	27.375 × 33.10	1.21	7.78	Coax fed probe, Narrowband
[22]	3.56 3.73	1	43.09 × 30.3	9.3	5.67	Coax fed, broadband
[23]	5.38	1	14 × 20	11.15	3.44	Coax fed, broadband
[24]	2.5,3.5, 4.5,5.6, 6.4,7.6, 9.2,9.9	8	38.14 × 47.6	8,5.7, 13,16, 10.6,7.9, 4.3,4	4.19	Edge fed, fractal, 19-slotted antenna
[25]	3.42 f_0=8 GHz	1	15.5 × 12.3	7.91	5.65	Edge fed, miniaturized, narrowband
[13]	2.45	1	66.4 × 66.4	49	7.16	Edge fed, UWB, high gain
This work	3.40 6.02, 8.21	3	28.45 × 33.4	30.88 24.60 4.50	3.59	Miniaturized, edge plus meander line fed, UWB

9.4 CONCLUSION

The antenna presented in this work is an example of an ANA with excellent impedance matching, enhanced gain, and radiation efficiency. Two numbers of wideband have been achieved at the resonating frequency by the introduced meander line. The meander line provides an artificial magnetic conductive (AMC) effect that is an alternative to the SRR. The radiation patterns of the antenna are omnidirectional. An Electric LC equivalent circuit has been generated for the proposed antenna. There are a number of possibilities to work with other combinations of Alphabets with Numerals. In the near future, there are many additional possibilities that the ANA can easily be made frequency/pattern/polarization/beam streaming reconfigurable by switching the ANA with a PIN diode and/or with a varactor diode placed at suitable places. Since, the antenna is simple, small, and comprises numbers and alphabets; therefore, the ANA structure can easily be used with the flexible substrate to play a role in the field of the wearable antenna on readymade clothes like jeans, T-shirts and jackets, baby toys, etc. These antennas can also be used for search and rescue operations in disaster management. Since the tested antenna supports all WiFi-3, WiFi-4, and WiFi-5 along with good support for sub-6 GHz 5G wireless communication bands. Thus the proposed antenna can easily be enabled by IoT and Zigbee-like devices and, therefore, it is an excellent candidate for D2D, M2M, home automation, V2X, and other green communication signals transmission and receptions. Because the antenna is tri-band in nature and covers multiple ranges of applications. Thus in the future 5G enabled smart cities network and device connections require such miniaturized size antennas with good gain and multi-purpose facility under the roof of a single umbrella.

KEYWORDS

- **DGS**
- **electric LC (ELC)**
- **meandered line**
- **miniaturized antenna**
- **Wi-Fi**
- **WiMAX and 5G wireless applications**

REFERENCES

1. Narang, T.; Jain, S.; Microstrip Patch Antenna: A Historical Perspective of the Development. In: *Conference on Advances in Communication and Control Systems*; CAC2S, 2013.
2. Sabath, F.; Mokole, E. L.; Samaddar, S. N. Definition and Classification of Ultra-Wideband Signals and Devices. *Radio Sci. Bull.* **2005,** *313*, 12–26.
3. FCC First Report and Order on Ultra-Wideband Technology. Feb. **2002,** 1–2.
4. Sharma, N.; Sandhu, S. A.; Design of Omni-Directional Rectangular Slotted Patch Antenna for Wireless Applications. *Int. J. Electr. Eng.* **2018,** *11* (1), 87–97.
5. Sharma, R. Design of Trapezoidal Patch Antenna with Inverted and Non-Inverted V-Shape Slot. *Int. J. Eng. Res. App.* **2013,** *3* (5), 1744–1747.
6. Krishna, R. V. S. R.; Kumar, R.; Design and Investigations of a Microstrip fed Open V-Shape Slot Antenna for Wideband Dual Slant Polarization. *Eng. Sci. Technol. Int. J.* **2015,** *18* (4), 513–523. DOI:10.1016/j.jestch.2015.03.005.
7. Louis, E.; Frenzel; Louis, E. Printed-Circuit-Board Antennas; 31 March **2005**; pp 1–2.

8. Yousif, B.; Sadiq, M.; Abdelrazzak, M. Design and Simulation of Meander Line Antenna for LTE Band. *Int. J. Sci. Eng. Res.* **2015,** *6* (7), 841–848.
9. Huang, Z. *Application of an Inductor Model to Design and Analysis of Printed Meander Line Antennas*; Taiwan, China **2007**; pp 10–12.
10. Leon, G.; Boix, R. R.; Medina, F. Comparison Among Different Reduced-Size Resonant Microstrip Patch. *Microwave Opt. Technol. Lett.* **2001,** *29* (3), 143–146.
11. Wang, Q.; Ning, M.; Wang, L.; Liu, J.; Wang, Y. Miniaturization Microstrip Antenna Design Based on Artificial Electromagnetic Structure. *Sixth Asia-Pacific Conference on Antennas and Propagation (APCAP)*; **2017**; pp 16–19. DOI: 10.1109/APCAP.2017.8420309
12. Bird, T. S. Definition and Misuse of Return Loss [Report of the Transactions Editor-in-Chief]. *IEEE Antennas and Propagation Magazine* **2009,** *51* (2), 166–167. DOI: 10.1109/MAP.2009.5162049.
13. Varshney, A.; Cholake, N.; Sharma, V. Low-Cost ELC-UWB Fan-Shaped Antenna Using Parasitic SRR Triplet for ISM Band and PCS applications. *Int. J. of Electron. Lett.*, **2021**. DOI: 10.1080/21681724.2021.1966655
14. Varshney, A.; Sharma, V.; Arya, V. Tri-Blade Table Fan Shaped Ultra-Wideband Microstrip Antenna Using Parasitic SRR Triplet. *Australian Patent*, **2021**, 2021101898.
15. Balanis, C. A. *Antenna Theory-Analysis and Design*; Vol. 2; John Wiley and Sons; 2017; pp 727–752.
16. Singh, S. P.; Singh, A.; Upadhyay, D.; Pal, S. K.; Munde, M. Design and Fabrication of Microstrip Patch Antenna at 2.4 GHz for WLAN Application Using HFSS. *J. Electron. Commun. Eng. (IOSR–JECE)*, **2021,** *1* (1), 1–6. DOI: 10.9790/2834-150100106
17. Ahmed, A. A.; Islam, S. M.; Masud, M. A.; Khan, N. H.; Zavala, B. J. W. A.; Islam, M. M. U. Design and Performance Analysis of 3.4 GHz Rectangular Microstrip Patch Antenna for Wireless Communication Systems. *Int. J. Commun. Antenna Propagation* **2017,** *7* (1), 80–86. DOI: 10.15866/irecap. v7i1.11423
18. Mansouri; Zahra; Afsaneh, S. A; Heydari, S.; Ferdows, B. Z. Dual Notch UWB Fork Monopole Antenna with CRLH Metamaterial Load. *Progress Electromagn. Res.* **2016,** *65*, 111–119.
19. Li, K.; Zhu, C.; Li, L.; Cai, Y.-M.; Liang, C.-H. Design of Electrically Small Metamaterial Antenna with ELC and EBG Loading. *IEEE Antennas Wireless Propagation Lett.* **2013,** *12*, 678–681.
20. Pozar, D. M. *Microwave Engineering*, Vol. 4; John Willey and Sons, 2012; pp 368–379.
21. Choudhury, D. R. Networks and Systems. *New Age Int. Pub.* **1998,** *1*, 123–135.
22. Sekra, P.; Dube, M.; Shekhawat, S.; Bhatnagar, D.; Saxena, V. K.; Saini, J. S. Broadband Rectangular Patch Antenna with Orthogonal Crossed Slits. *Int. J. Microwave Opt. Technol.* **2011,** *6* (4), 179–184.
23. Kiroriwal, M.; Rawat, S. Improvement in Radiation Parameters of Rectangular Microstrip Patch Antenna. *Int. J. Eng. Res. Gen. Sci.* **2014,** *2* (6), 393–397.
24. Prasad, G. G.; Kumar, G. S. Design of Enhanced Gain Multiband Miniaturized Antenna for Mobile Communications. *Turk. J. Comput. Math. Educ.* **2021,** *12* (11), 4719–4729.
25. Kiani, S. H.; Imran; Munir M.; Abdullah, M. A.; High Miniaturaized Antenna for Wi-MAX and Small Wireless Technologies. *J. Mech. Cont. Math. Sci.;* **2019,** *14* (1), 250–257. DOI: 10.26782/jmcms.2019.02.00017.

26. Tang, L.; Li, J.; Wang, Y.; Zhao, L.; Jiang, Q. Adsorption Capability for Congo Red on Nanocrystal HneMFe204 (M=Mn, Fe, Co, Ni) spinel ferrites. *Chem. Eng. J.* **2010.** DOI: 10.1016/j.cej.-2011.10.088
27. Pradhan, D.; Tun, H. M. Circular-MSPA: Design and Analysis of Applications Intended for 5G Environment. *J. Netw. Secur. Comput. Netw.* **2023,** *9* (1), 14–19. https://doi.org/10.46610/jonscn.2023.v09i01.002
28. Pradhan, D.; Dash, A.; Tun, H. M.; Wah, N. K. S.; Oo, T. A Sustainable Key Enabler for mm-Wave Beamforming in 5G Environment. *J. VLSI Design Sign. Process.* **2022,** *8* (3), 10–17. https://doi.org/10.46610/jovdsp.2022.v08i03.002
29. Pradhan, D.; Ray, S.; Dash, A. A Critical Review on Sustainable Development of Green Smart Cities (GSCs) for Urbanization. *J. Altern. Renew. Energy Sources* **2022,** *8* (3), 21–29. https://doi.org/10.46610/joares.2022.v08i03.003
30. Dash, A.; Pradhan, D.; Tun, H. M.; Naing, Z. M. (2022). m-MTC for Optimized Communicationin5G.*J. Netw. Secur. Comput. Netw.* *8*(3), 1–8. https://doi.org/10.46610/jonscn.2022.v08i03.001

CHAPTER 10

5G-Enabed Smart Healthcare System with the Integration of Blockchain Technology

HLA MYO TUN[1], DEVASIS PRADHAN[2], ZAW MIN NAING[3], THANDAR OO[4], and NAW KHU SAY WAH[4]

[1]*Research Department, Yangon Technological University, Yangon, Myanmar*

[2]*Department of Electronics & Communication Engineering, Acharya Institute of Technology, Dr. Sarvepalli Radha Krishnan Road, Bengaluru, India*

[3]*Department of Research and Innovation, Ministry of Science and Technology, Yangon, Myanmar*

[4]*Department of Electronic Engineering, Yangon Technological University, Yangon, Myanmar*

ABSTRACT

The chapter presents the 5G-Enabed Smart Healthcare System with the Integration of Blockchain Technology. The challenging issues of smart healthcare system and security with blockchain technology in 5G era and beyond are crucial problems nowadays. The first portion is the understanding on blockchain technology and 5G telecommunication system and beyond. The second portion of this study is the linkage between blockchain schemes and smart healthcare systems. The third portion is

5G Green Communication Networks for Smart Cities. Devasis Pradhan, Rajeswari, Hla Myo Tun, Naw Khu Say Wah, & Thandar Oo (Eds.)

the implementation of the overall system design for a complete system. The ideas from the society needs for smart healthcare system based on 5G system and beyond are utilized and solved the above mentioned problem statements with blockchain technology.

10.1 INTRODUCTION

The blockchain that one a file-a collective and public ledger of transactions that records all transactions from the creation block (first block) in anticipation of the present day. At this juncture, records are put in safekeeping in blocks and those blocks are cryptographically chained together. The basic operation of a blockchain consists of (i) transaction, (ii) validation, (iii) broadcast valid transactions, (iv) consensus on the next block creation, and (v) chaining blocks to form an immutable record. Figure 10.1 shows the Blockchain Concepts in Healthcare System.

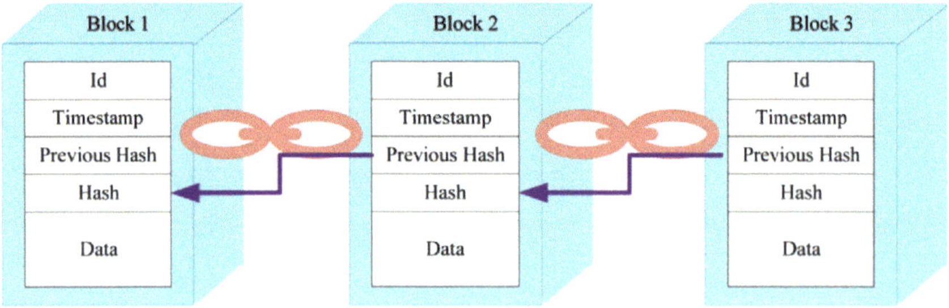

FIGURE 10.1 Blockchain concepts in healthcare system.[1]

Transactions are the records that are put in safekeeping in a blockchain. When somebody sends a transaction, the software that is used in bank will use the private key of that person to cryptographically sign the transaction for the confirmation. That signature would prove ownership of the unspent output for the process and authorize undertaking of the money. Therefore, when a node catches a new transaction, it checks to make sure that the signature is valid or not. If the signature is not valid, it will pay no attention to the transaction. In broadcast valid transactions, the processes are very systematic. All valid transactions would be broadcast to the peer nodes in the blockchain network rather than the centralized "Hub and Spoke" type of blockchain network, Blockchain is a decentralized or spread out peer-to-peer network where each node has a duplicate of the ledger. The consensus algorithm confirms that the next block in a blockchain is the one and only version of the truth. It keeps influential supporters from tremendous the system and efficaciously forking the chain in the blockchain network. Many consensus mechanisms have pros and cons for algorithm creation. Once the block is added to the chain in the blockchain network, the transactions that block become immutable, or impossible to remove. The consensus models consist of (i) Proof of Work, (ii) Proof of Stake, (iii) Practical byzantine fault tolerance, (iv) Delegated proof of stake, (v) Round robin consensus model, (vi) Proof of authority (identity) model, and (vii) Proof of elapsed time (PoET) consensus model. In technical terms for chaining blocks to form an immutable record, blockchains are written and read-only processes in the network. All blocks are usually encrypted and use standard encryption practices for chaining blocks. Some blockchains agree to "BYOE" (Bring Your Own Encryption) scheme.[2]

10.2 ROLE OF BLOCKCHAIN IN HEALTHCARE

Healthcare is an outstanding industry that is presently being transformed using the up-to-the-minute technology, consequently, it could meet the challenges it is facing in the 21st century. Complex Technology can benefit healthcare organizations to meet growing demand and resource-fully operate to convey better healthcare processes for all patients.

10.2.1 CHALLENGES AND SOLUTIONS IN HEALTHCARE SYSTEM

The challenges in Healthcare Systems are (i) the burden of preventable medical errors, (ii) medical information explosion, (iii) the slow diffusion of medical knowledge, (iv) the high elderly population, (v) the high cost of hospitalized treatments, and (vi) quality management of drugs. The solution for challenges in healthcare system is the blockchain with smart contracts. It is one of the most exploited applications of blockchain technology in the present times. The impressions of ideas on smart contracts were familiarized by Nick Szabo in 1994. It is a set of computer codes between two or more parties that run on the top of a blockchain system. If one of the pre-defined rules is met; the smart contract accomplishes that to produce the output from the blockchain network.[1]

10.2.2 FACTORS AFFECTING TO HEALTHCARE FROM BLOCKCHAIN TECHNOLOGY

There are many factors affecting healthcare from blockchain technology. The following are the important factors. They are (i) trust building, (ii) data exchange acceleration, (iii) lower costs, (iv) security and privacy improvement, (v) accountability and immutability, and (vi) automation via smart contracts.[1]

10.3 5G HEALTHCARE

As 5G technology delivers an exact high internet connection speed and necessitates less time to receive or send big data, it is essentially suitable for the healthcare division. Healthcare facilities are a very significant and indispensable requisite of humanity. With the initiation of novel technologies in the healthcare division, it has become easier to reconcile patients suffering from critical diseases.

The unadventurous healthcare management system is actually complicated and necessities to be advanced with the intention of proper care could be engaged of all patients as per their necessities. The up-to-the-minute technology assistance to be responsible for treatment facilities by means of a wireless telecommunication system, and if the number of patients increases, there is certainly for a medical doctor. He or she could accomplish his or her time from home for the reason that there is no need to go anywhere, and so it also saves time and money.

When applied in the clinical areas, 5G technology with IoT confirmations implausible perspective to empower reserved medical techniques. In the present circumstances, remote surgery is precise in effect for patients who are far-flung from the city or stay in remote whereabouts and also for patients with mental infirmities, depression teething troubles, children with autism, sleep-deprived patients, and so on.

5G technology could affect healthcare system with the following important applications. They are: (i) telemedicine, (ii) monitoring of patients living in remote areas, (iii) augmented and virtual reality, (iv) analysis of data, (v) decentralization of the ideal healthcare model, and (vi) transfer of large files.[3,6–8]

10.4 INTEGRATION OF BLOCKCHAIN, 5G, AND HEALTHCARE

The fantasy of smart hospitals could be accomplished after the creation of 5G because it could be responsible for high internet bandwidth and speed. To be supplementary explicit, 5G could open novel opportunities for patient care while consuming machine learning and AI after blockchain is encompassed. With 5G, the perspective of the use of the internet with security would increase. When exhausting smart hospitals, the highest problem is retaining the data of all patients and providing security. Several private organizations arrange for services to maintain hospital records and data, but they do not guarantee data security for all patients.

The blockchain technology is intended with high-performance software, which can integrate and accomplish many contracts, whereas Bitcoin cannot be utilized for big data. A blockchain[9] scheme with 5G upkeep could associate the healthcare providers and be responsible for dependable data sharing. It is correspondingly suitable for patients for supervising their health history and if compulsory, for sharing their medical data with healthcare specialists. Subsequently, data security is very high, the healthcare benefactor can acquire precise and accurate patient data. In that circumstance, the patient could get appropriate treatment in time.

The wireless cell era transformed into inaugurated around 1970 with the establishment of the cell community, bringing up to date the old-fashioned machine, and emergent novel generation configurations, as shown in Figure 10.2.[4]

Technologies used in the 5G network system are organized as follows: They are: (i) millimeter waves, (ii) small cells, (iii) Big multiple input/output system, (iv) beamforming, (v) full duplex, and (vi) software-defined networks.

The advantages of the 5G networking system with the solicitation of IoT are recorded as follows:

- It affords high quality performance.
- It offers connectivity to a large number of devices efficiently.

- It improves the operating efficiency of the system and supports high-quality videos utilized in online healthcare services.
- It provisions high bandwidth for communication services.

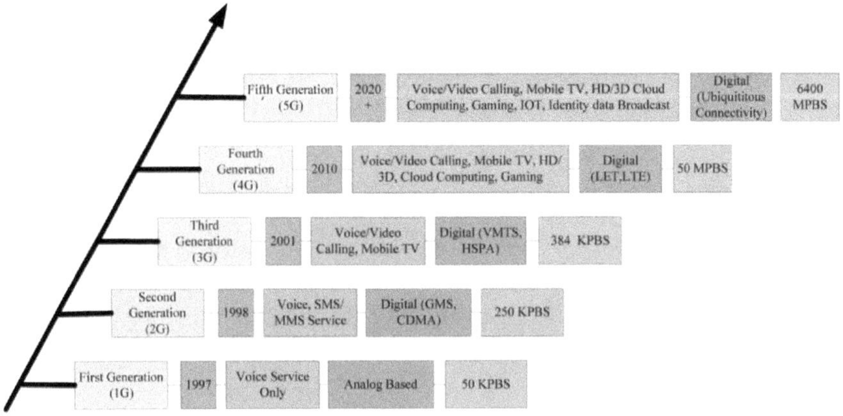

FIGURE 10.2 Evolution from 1G to 5G.

Source: Adapted from Ref.[4]

The detailed processes for integration of 5G technology and healthcare systems are based on the following fields. They are (i) Smart clothing, (ii) Diagnosis services in rural areas, (iii) Management in hospitals, (iv) Use of robots, (v) Monitoring of healthcare data, (vi) Imaging, (vii) Diagnostics, and (viii) Data analytics and treatment.

The impact of 5G on medical access, quality, and cost is based on the merits of the integration. 5G technologies are accessible with high-speed connectivity, helping to share big data and high-definition video calling, which is very appropriate for the digital medicine system. Video conferencing techniques are very efficient for difficult conditions of patients face to face. If they use digital medical services, they could save time and transportation charges. 5G-enabled smart devices benefit doctors by getting accurate reports on time, assessing the patient details at the same time, and contributing a suitable solution. So there is no risk of wrong diagnosis of diseases, and patients acquire correct information about the progress of their health.

5G techniques play a crucial role in telemedicine services and could be more in effect by utilizing the following steps such as: (i) continuous monitoring, (ii) predictive analytics, (iii) impact on business models, (iv) remote diagnosis and imaging, and (v) improved state of art.

10.5 RESEARCH TRENDS FOR 5G HEATHCARE SYSTEM WITH BLOCKCHAIN

The main use cases for blockchain in healthcare system are based on the specific approaches with Patient Data Management Blockchain Approaches, Blockchain-based Efficient Electronic Health Records Access Approaches, Blockchain-based Tracking Clinical Trials Approaches, and Stopping Counterfeit Drugs Blockchain Approaches.

The records for human medical lean toward being divided by agencies of health service making it intolerable to regulate a medical history of patient without looking up their care provider in the aforementioned. That progression could earnings a noteworthy expanse of time, and may time and again consequence in inaccuracies due to error of human. Blockchain could be utilized to accumulate all of the information of patients in one place, making it unpretentious for patients and doctors to observe. In the aforementioned contemporary design, benefactors sustain the blockchain over and done with the Proof of Authority (PoA) contrivance.

The electronic health records (EHR) could be inflexible to accomplish. One provider of healthcare system EHR for a patient may contrast with the other provider for the same patient. Blockchain could position the control in the hands of the patient. All patient has complete right of entry and switch over their individual personal health documents. A patient would have the capability to countenance the handover of health records from one medical doctor to another.

Numerous stakeholders convoluted in the clinical trial progression from the patients to the benefactors to the supervisory body, and others, varieties for an intricate categorization of information allotment. Ahead of that, the environment of clinical trial information, which is complex and comprehends the medical information of patients supplements a superfluous impediment. Nevertheless, the information must be threatened, it also must be collected blatantly between individuals complicated in the clinical trial. Blockchain-based clinical trial way out would intensification the speediness of drug improvement and by this means subordinate the charge and approachability of first-hand medicines in the market.

Observance of a forgery-proof record of relations on a blockchain facilitates to substantiate raw resources that drugs are completed of and perceive counterfeits at the appointed time.

10.6 CONCLUSION

5G is responsible for a precise high internet connectivity speed for the users, controls robotics in real time, transfers big data with high-speed condition, and has several applications used in healthcare areas. 5G supports the high-speed mobile system, which progresses healthcare properties in a wide range of situation. The 5G wireless networking system with its solicitation could transmute healthcare by enhancing human capacity, transport, and healthcare paraphernalia in rural and remote areas in real time applications. It consents to operating virtualization, a high-quality telemedicine scheme that provides high performance and dependability. It cannot tolerate illegal users accessing the networking system using high security and privacy. Operating such a high-speed system involves a high-security system to protect the important hospital data and personal records of all patients. 5G-operated mobile systems facilitate the personalization of healthcare services. The personalization of medical services assistances in contributing supplementary precaution, more care, proper monitoring, and provides less time to progress health of patients, and furthermost expressively decreases the cost of treatment.

KEYWORDS

- **5G telecommunication system**
- **smart healthcare system**
- **blockchain technology**
- **system implementation**

REFERENCES

1. Liyanage, M. Blockchain in Healthcare, 2010. 10.13140/RG.2.2.13096.19205.
2. Thangamuthu, P.; Ranganathan, I.; Mani, K.; Shanmugam, S.; Palanimuthu, S. *Chapter 1—Blockchain Technology and Its Relevance in Healthcare in Blockchain and Machine Learning for e-Healthcare Systems*; The Institution of Engineering and Technology: London, 2020.
3. Tanwar, S. *Blockchain for 5G Healthcare Applications Security and Privacy Solutions*; The Institution of Engineering and Technology: London, 2021.
4. Hwang, K.; Min, C. *Big Data Analytics for Cloud, IoT and Cognitive Computing*; John Wiley & Sons: New York, 2017, pp 234–267.
5. Tun, H. M. Photoplethysmography (PPG) Scheming System Based on Finite Impulse Response (FIR) Filter Design in Biomedical Applications. *Int. J. Electr. Electron. Eng. Telecommun.* **2021,** *10* (4), 272–282. Doi: 10.18178/ijeetc.10.4.272-282.
6. Pradhan, D.; Sahu, P. K.; Dash, A.; Tun, H. M. Sustainability of 5G Green Network toward D2D Communication with RF- Energy Techniques. In: *2021 International Conference on Intelligent Technologies (CONIT)*; **2021**; pp 110. DOI: 10.1109/CONIT51480.2021.9498298
7. Pradhan, D.; Sahu, P. K.; Ghonge, M. M.; Rajeswari, Tun, H. M. Security Approaches to SDN-Based Ad hoc Wireless Network Toward 5G Communication. In: *Software Defined Networking for Ad Hoc Networks*; Ghonge, M. M.; Pramanik, S.; Potgantwar, A. D., Eds.; EAI/Springer Innovations in Communication and Computing: Springer, Cham. https://doi.org/10.1007/978-3-030-91149-2_7
8. Pradhan, D.; Sahu, P. K.; Rajeswari, Tun, H. M. A Study of Localization in 5G Green Network (5G-GN) for Futuristic Cellular Communication. In: *Proceedings of the 3rd International Conference on Communication, Devices and Computing. Lecture Notes in Electrical Engineering*; Sikdar, B.; Prasad Maity, S.; Samanta, J.; Roy, A., Eds., Vol. 851; Springer: Singapore, 2022. https://doi.org/10.1007/978-981-16-9154-6_43
9. Pradhan, D.; Agarwal, A.; Tun, H. M.; Naing, Z. M.; Oo, T. Critical Security & Privacy Issue in Blockchain Technology Intended to Industry 4.0. *Middle East Res J. Eng. Technol.* **2022,** *2* (1), 1–7.
10. Pradhan, D.; Tun, H. M. Security Challenges: M2M Communication in IoT. *J. Electr. Eng. Autom.* **2022,** *4* (3), 187–199.
11. Pradhan, D.; Sahu, P. K.; Goje, N. S.; Ghonge, M. M.; Tun, H. M.; Rajeswari, R.; Pramanik, S. Security, Privacy, Risk, and Safety Toward 5G Green Network (5G-GN). *Cyber Secur. Netw. Secur.* **2022,** 193–216.
12. Pradhan, D.; Priyanka, K. C. RF-Energy Harvesting (RF-EH) for Sustainable Ultra Dense Green Network (SUDGN) in 5G Green Communication. *Saudi J. Eng. Technol.* **2020,** *5* (6), 258–264.
13. Pradhan, D.; Tun, H. M.; Dash, A. K. IoT: Security & Challenges of 5G Network in Smart Cities. *Asian J. Convergence Technol. (AJCT)* **2022,** *8* (2), 45–50.
14. Pradhan, D.; Sahu, P. K.; Ghonge, M. M.; Tun, H. M. Security Approaches to SDN-Based Ad hoc Wireless Network Toward 5G Communication. In: *Software Defined Networking for Ad Hoc Networks*; Springer International Publishing: Cham, 2022; pp. 141–156.

15. Pradhan, D.; Sahu, P. K.; Tun, H. M. A Study of Localization in 5G Green Network (5G-GN) for Futuristic Cellular Communication. In: *Proceedings of the 3rd International Conference on Communication, Devices and Computing: ICCDC 2021*; Springer: Singapore, 2022; pp 453–465.
16. Wah, N. K. S. Integration of AI/Ml in 5G Technology Toward Intelligent Connectivity, Security, and Challenges. *Mach. Learn. Algorithms Appl. Eng.* **2023,** *239*.
17. Priyanka, K.; Mallavaram, G.; Raj, A.; Pradhan, D. Cognitiveness of 5G Technology Toward Sustainable Development of Smart Cities. *Decis. Supp. Syst. Smart City App.* **2022,** 189–203.
18. Sinha, H. K.; Kumar, A.; Saurabh, A.; Pradhan, D. Comparative Analysis of Papr Using PTS and SLM Reduction Techniques for 5G Communication System. *Asian J. Converg. Technol. (AJCT)* **2022,** *8* (2), 51–56.
19. Singh, H. W.; Devaki, K.; Fernandes, J. V.; Pradhan, D. PoD Vs SNR Estimation: C-MIMO Radar Using STC and STAP Algorithm. In: *2022 IEEE Region 10 Symposium (TENSYMP)*; IEEE, 2022; pp 1–6.
20. Patil, P.; Pawar, P. R.; Jain, P. P.; KV, M.; Pradhan, D. (2020). Performance Analysis of Energy Detection Method in Spectrum Sensing Using Static & Variable Threshold Level for 3G/4G/VoLTE. *Saudi J. Eng. Technol.* **2020,** *5* (4), 173–178.

CHAPTER 11

Integration of IoT with Blockchain Technology (BIoT) for Data Management and Security Toward Remote Healthcare System (RHS)

DARA JOGESH KUMAR[1], NITIN S GOJE[2], DEBASIS DAS[2], DEVASIS PRADHAN[3], HLA MYO TUN[4], and ALESSANDRO BRUNO[4]

[1]*Software Developer, HTC Global Services, Begumpet, Hyderabad, India*

[2]*Department of MIS, Webster University, Tashkent, Uzbekistan*

[3]*Department of ECE, Acharya Institute of Technology, Bangalore, India*

[4]*Rector, Research Department, Yangon Technological University, Myanmar*

ABSTRACT

The turn of events and reception of Electronic Health Records (EHR) and health monitoring Internet of Things (IoT) gadgets have empowered the digitization of patient records and have likewise considerably changed the medical services conveyance framework in perspectives, for example, far-off understanding of observing medical services direction, and clinical exploration. In any case, information will generally be divided among well-being frameworks and forestall clinical data interoperability at the point of care. Toward network deployment with versatility and intrinsically having dormancy in exchange handling in blockchain technology, there is space for investigating and growing new strategies to use the security

5G Green Communication Networks for Smart Cities. Devasis Pradhan, Rajeswari, Hla Myo Tun, Naw Khu Say Wah, & Thandar Oo (Eds.)

elements of blockchains inside medical services applications. In any case, IoT-based medical services observing frameworks have issues connected with security breaks and protection of medical care information. In this manner, planning some security conventions for IoT-based medical services following frameworks is crucial. Every once in a while, analysts propose various traditions to get information on IoT-based medical services following frameworks. We give the subtleties of different potential models of BIoT-RHS. We additionally provide the subtleties of other dangers and assaults alongside the danger model in BIoT-RHS. The security necessities and uses of BIoT-RHS are likewise examined. We then, at that point, give the outline of different existing security conventions connected with BIoT-RHS

11.1 INTRODUCTION

The improvement of blockchain innovation depends on an assortment of disciplines, including cryptography, math, calculations, and financial models. All cryptographic money exchanges are recorded on a computerized and decentralized. Clients might monitor their crypto exchanges by looking at an ordered rundown instead of a unified narrative. The blockchain's application potential is splendid and has previously delivered results since its start. In different fields, blockchain innovation has been integrated and sent with trendy shrewd agreements from the earliest long periods of cryptographic forms of money to the current day. Uprightness, secrecy, and accessibility are simple ideas used to portray security.

However, their honesty and accessibility are ensured by these frameworks. In these blockchain frameworks, information availability is generally of better quality. Information replication for appropriated frameworks increments lucid information accessibility but diminishes compose accessibility. Despite the way that the blockchain's hidden engineering is incredibly protected, executions of state-of-the-art advancements have taken utilization of the blockchain's security highlights. Since all of the organization's public keys are apparent to everybody, the blockchain framework is powerless to the spilling of conditional security. An assistive innovation survey method is encouraged to accomplish the most significant level of independence in individuals with hindrances. The assessment system ought to consider both the singular's necessities and assets and the hardware's advantages and drawbacks.

11.2 RELATED WORKS

Kevin Ashton initially instituted the IoT innovation in 1999 to integrate omnipresent sensors advancing the Radio Frequency Identification innovation to work with implanted Internet. According to the insights, in 2020, around 127 new IoT-based gadgets are being sent with the Web for organization and cloud-based activities. According to perspective specialists, it is assessed that approximately 31 billion IoT-based devices are being used, and 35 billion IoT gadgets to be introduced by 2021. The wake of the computerized upset has brought about an enormous expansion in the new IoT gadgets, advancing more than 75 billion new devices toward the finish of 2025. The IoT innovation has worked over the long run and has re-imagined the correspondence among the hubs in the organizations, which has improved the extent of IoT economically and robotized administrations. However, the enormous number of clients and hubs in the organizations and the moving of information have fundamental issues connected with the information's security and protection, as examined in Tawalbeh, L. et al.,[2] Srinivasu, P.N. et al.[3]; Naga Srinivasu et al.[4]

Blockchain innovation is decisive in guaranteeing the security of the information traded over IoT engineering.[13,14] In this way, the patient checking models and the medical services of the executive's system that depends on the IoT engineering for checking are driven by blockchain innovation, which can work over conveyed network models with more extensive security ascribed over cutting-edge cryptographic crude. The medical services of the executive's models, like Remote patient observing edges works,[15,16] are driven by the blockchain innovation that guarantees information security by saving the exchange history of the hubs in the engineering. Blockchain engineering is based on the verification of work rules. An exchange is considered authentic provided the organization approves that approved seats have performed adequate computational exertion. Along these lines, the innovation in light of the blockchain model is being created to guarantee approved admittance and security to patients' clinical records in such a unique circumstance.

11.3 IoT

It is a term used to describe a network of interconnected physical devices, appliances, vehicles, and other objects that are embedded with sensors, software, and internet connectivity, enabling them to collect and exchange data. The main idea behind IoT is to make devices smarter and more interconnected, allowing them to communicate and work together to perform tasks and enhance user experiences. For example, a smart thermostat can connect to the internet, gather information about the user's preferences and habits, and adjust the temperature of their home automatically.

IoT is being used in various industries and applications, including smart homes, wearable devices, healthcare, transportation, agriculture, and industrial automation. In smart homes, IoT devices can control lighting, temperature, and security, while wearable devices can monitor vital signs and track fitness data. In healthcare, IoT can be used to monitor patients remotely, track medication usage, and manage chronic diseases. In agriculture, IoT sensors can collect data about soil moisture, temperature, and other environmental factors, helping farmers to optimize crop yields. While IoT offers many benefits, it also comes with potential security risks, such as data breaches and unauthorized access. It is important to prioritize security measures, such as encryption, authentication, and access controls, to protect IoT devices and the data they collect.

11.4 CORE COMPONENT OF RHS

11.4.1 BLOCKCHAIN TECHNOLOGY AND RELATED CONCEPT

Blockchain is a decentralized, digital ledger that records transactions and data across a network of computers, making it difficult to alter or hack. Each block in the chain contains a unique cryptographic hash, a timestamp, and a record of transactions that have been verified and added to the chain. Once a block is added to the chain, it cannot be modified, creating a permanent and transparent record of all transactions.

Related concepts to blockchain include:

- *Cryptocurrency:* A digital currency that uses cryptography for security and operates independently of a central bank. Bitcoin is the most well-known cryptocurrency and is based on blockchain technology.
- *Smart contracts:* Self-executing contracts that are programmed to automatically execute when certain conditions are met. They are often used in blockchain-based applications to automate processes and reduce the need for intermediaries.

- *Distributed ledger:* A database that is spread across multiple locations, making it difficult to hack or modify. Blockchain is a type of distributed ledger.
- *Decentralized applications (dApps)*: Applications that run on a decentralized network and use blockchain technology to store and manage data.
- *Public vs. private blockchain:* Public blockchains are open to anyone, while private blockchains are restricted to a select group of users. Public blockchains are often used for cryptocurrencies, while private blockchains are used for enterprise applications.

11.4.2 IoT AND RELATED CONCEPTS

Related concepts to IoT include:

- *Sensors:* Sensors are a crucial component of IoT devices, allowing them to collect data and respond to their surroundings.
- *Edge computing:* A computing model that processes data closer to the source. Edge computing is often used in IoT applications to reduce latency and improve real-time processing.
- *Artificial intelligence (AI):* It is an intelligent way to perform AI is often used in conjunction with IoT for precise decision making with relevant details.
- *Cloud computing:* A model for delivering on-demand computing resources over the internet. Cloud computing is often used in IoT applications to handle huge amount of data.
- *Machine learning:* It is a part of AI that leads to training machines to predict the correct decision with suitable data analysis. Machine learning is often used in IoT applications to analyze data and improve performance.
- *Wearable devices:* Smartwatches and fitness trackers. Wearable devices are a popular application of IoT, collecting data about the user's health and activity levels.

IoT has many potential applications, including smart homes, healthcare, transportation, and industrial automation. However, it also raises concerns about security and privacy, as IoT devices can collect sensitive data about individuals and organizations. It is important to prioritize security measures, such as encryption and access controls, to protect IoT devices and the data they collect. The IoT is the process of bringing together disparate companies and devices. In order to make decisions about IoT centralized options, there must be a strong connection between the path and the things.[24] The detecting space and the organization area are corresponding to the IoT door. Advances like Zigbee, Bluetooth, and WiFi make it possible to connect Smart Things to IoT gateways. The IoT continuously equips a large number of devices and slots with a network. The IoT is used in everyday life for a variety of things that are connected to the Internet, such as online shopping, wearable technology, smartphones, automobiles, home lighting, appliances, and so on. IoT in clinical benefits: Medical care faces various issues and troubles that can be managed utilizing the IoT.[29] Furthermore, the clinical consideration capacities can be overhauled multifold utilizing the IoT. In the medical services region, there is a shortfall of steady data, sagacious card contraptions, wrong standard examination, and various overhauls like remote checking of patients that can be made possible using IoT.

11.4.3 INTEGRATION OF BLOCKCHAIN AND IOT TECHNOLOGIES IN RHS

RHS could benefit from the integration of blockchain and IoT technologies. By combining these technologies, RHS can improve the security, transparency, and efficiency of healthcare data. Blockchain can provide a platform for data security and feasible access to authenticate person in the health care industry with the consent of stakeholder. Blockchain can also facilitate the secure sharing of data among healthcare providers and patients, improving collaboration and coordination of care.

IoT can provide real-time monitoring of patient's health and provide data for analysis and diagnosis. By connecting various medical devices, wearables, and sensors, IoT can collect and transmit data to healthcare providers, allowing them to monitor patients remotely and provide early intervention when necessary. With the help of machine learning algorithms, IoT can also identify patterns and anomalies in health data, enabling better diagnosis and treatment.

By integrating blockchain and IoT technologies, RHS can improve data security, privacy, and interoperability. Patients can have more control over their health data, allowing them to share it securely with healthcare providers and researchers. A complete report of the patient's health history is provided by medical institution, enabling better diagnosis and treatment. Researchers can access large sets of data for analysis and development of new treatments. One example of this integration is MedRec, a blockchain-based electronic medical record system that uses IoT sensors to collect and transmit patient data. MedRec allows patients to control their health data and share it securely with healthcare providers. The system is designed to improve patient privacy, reduce fraud, and enable more efficient healthcare delivery.

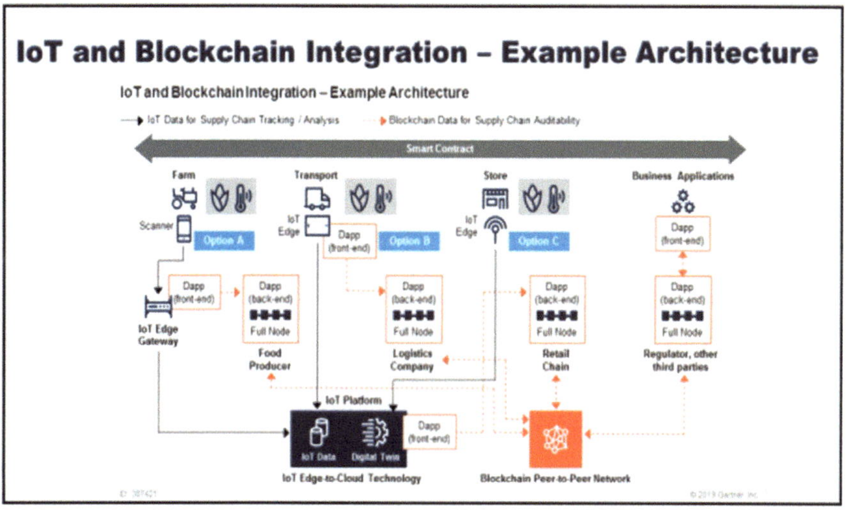

FIGURE 11.1 Integration of IoT with blockchain for RHS

Source: Adapted from blogs.gartner.com

Overall, the integration of blockchain and IoT technologies in RHS. However, it is important to address the challenges of data interoperability, standardization, and scalability to ensure widespread adoption and successful implementation.

11.4.4 ARCHITECTURE OF RHS

Specific medical services applications need the immediate consideration of clinical representatives, like in illnesses connected with the heart or when giving medical services to patients met with hazardous mishaps. Such circumstances require a constant and essential reaction with essentially less inertness. In a general cloud climate, the inactivity for information to be communicated to the cloud, handling in the cloud, and getting a reaction includes huge idleness, which is not adequate. To survive or restrict the dormancy issues, we can utilize haze processing, which carries the registering and stockpiling assets to the edge of the organization, that is, nearer to the sensors. The more significant part of the current medical care arrangements utilizes a cloud climate for navigation. Lately, an ever-increasing number of proposed arrangements are considering mist processing for time-basic medical care applications.

The detecting layer contains various sensors for checking the well-being vitals of patients and old individuals. These sensors send the information to local application gadgets or equipment stages like mobiles, Arduino, Raspberry Pi, etc. These application gadgets, this way, forward the crude information to the following layer, the haze layer. The mist layer contains a neighborhood door and at least one transient stockpiling server where information can be sifted or investigated in light of the application prerequisites. Security and protection strategies can likewise be executed at the nearby passage. After the necessary information changes, it is sent to a line entryway, which will advance the information to the cloud layer. A firewall can be put at the neighborhood or boundary door to channel pernicious traffic. We accept that the correspondence between elements in the organization is brought out through secure channels. At the cloud layer, the information can be put away on the highly durable capacity servers, or it tends to be examined for decision making by the clinical workforce.

11.5 CHALLENGES OF USING BLOCKCHAIN IN HEALTHCARE-DERIVED INDUSTRIAL IOT

Using blockchain technology in healthcare-derived industrial IoT (IIoT) applications:

- *Scalability:* A major aspect of blockchain technology is its limited scalability. Due to numerous transactions, the blockchain network makes it increasingly difficult to process and store data. This can be particularly challenging in IIoT applications.
- *Interoperability:* Blockchain technology requires standardization and interoperability between different systems in order to function effectively. In the healthcare-derived IIoT space, there are often many different systems and devices from different vendors, making interoperability a challenge.
- *Security:* While blockchain technology is generally considered to be secure, there are still concerns around the security of the

devices that are connected to the network. In healthcare-derived IIoT applications, where the security of patient data is of paramount importance, ensuring the security of these devices is critical.

- *Regulation:* The healthcare industry is heavily regulated, and blockchain technology presents new challenges in terms of compliance. Regulations around data privacy and ownership, for example, may need to be reevaluated and updated in order to accommodate the use of IIoT with the driving force of blockchain.
- *Cost:* Implementing blockchain technology can be costly, both in terms of infrastructure and development.
- *Accuracy and Reliability:* RHMS needs elemental precision and unwavering quality of medical services data. The precision and steadfast quality systems should be guaranteed over ordinary usage because deceptive data could become temperamental and problematic to patients. The IoT innovation has a powerful personal stake, for instance, to consolidate and decipher various clinical data in proficient dynamic cycles in far-off medical services observing frameworks. Through this, doctors will examine the condition of each patient's well-being, and subsequently, more compelling intercessions should be possible.

Overall, while blockchain technology holds great promise for healthcare derived IIoT applications, there are several challenges that must be addressed in order to fully realize its potential. These challenges will require collaboration between industry stakeholders, including healthcare providers, technology vendors, and regulatory bodies, in order to ensure that the benefits of this technology can be realized while mitigating its risks.

11.6 USED CASES AND PRACTICAL SCENARIO

11.6.1 IoT DEVICES

IoT devices are physical devices that are connected to the internet and are capable of sending and receiving data. IoT devices are designed to be connected and communicative, which means that they can work together to perform complex tasks and share information.

For example, a smart home system could use data from a variety of devices, including sensors, cameras, and appliances, to provide users with insights and control over their home's security, energy usage, and environmental conditions. However, these interconnections also mean that

IoT devices can be vulnerable to cyberattacks and security breaches, highlighting the importance of strong security measures and privacy protections.

11.6.2 IoT ACCESS CONTROL

In the context of using blockchain technology in healthcare, access control mechanisms are critical to ensuring the security and privacy of patient data stored on the blockchain. Some access control mechanisms that can be used in this context include:

- *Identity management:* Blockchain-based identity management can be used to authenticate users and devices before granting access to the blockchain network. This can involve using digital signatures or cryptographic keys to verify identity.
- *Smart contracts:* Smart contracts control policies in an autonomous manner and enforce them on the blockchain. This can include defining access rights and permissions based on pre-defined rules and conditions.
- *Consensus mechanisms:* Consensus mechanisms such as proof of stake or proof of authority can be used to limit access to the blockchain network to only authorized users or devices.
- *Private blockchain networks:* Private blockchain networks can be used to restrict access to the blockchain to only authorized users or devices. This can be especially important in healthcare where patient data must be kept private and secure.
- *Data encryption:* Data encryption can be used to protect sensitive patient data stored on the blockchain. This involves encoding data in a way that can only be deciphered by authorized users or devices.

11.6.3 DATA INTEGRITY AND SECRECY

In a disseminated record worldview, it is frequently trusted that information trade while keeping up with adequate privacy will be entirely possible.[60] Therefore, the capacity to hold the arrangement of advanced marks and information hashes given by BC might be utilized to successfully attest information trustworthiness and IoT-related information. A utilization case for this can be as per the following. Coming up next are a few instances of potential applications.

A caution framework for a home or working environment might be overseen by an assortment of individuals with changing degrees of access certifications. Assuming that interlopers gain admittance to it, policing may have to utilize remote admittance to examine. A disseminated ledger, including many interconnected gadgets, may be especially advantageous in this circumstance

A unique medical care dart needs might be imparted to the scientist or clinical staff by utilization of an individual wellness tracker. Subsequently, an individual might be prepared to pay a small premium for administrations given by a producer. At the point when savvy houses include weather conditions station/air observing IoT items that are shared by many gatherings, a similar circumstance could happen. The circulated record might be the leading choice for an organization of machine makers, professionals, and specialists with all the preposterously huge earmarks.

11.6.4 SMART CONTACTS

There are a few circumstances, including different associations, in which it is critical to decide if every one of them is consented to. In this manner, BC's brilliant agreements might be utilized rapidly and consistently. Coming up next are a few instances of potential applications. For example, conveyed records can be used by certain people who share individual information with their medical care supplier to guarantee that the central approved clinical staff approaches the data. Preferably, the drug store and the overall professional in a multi-party framework should convey the patient's circulatory strain readings to work with the simple administering of a suggested prescription.

11.7 CASE STUDY: IOT-BASED HEALTHCARE SYSTEM

The IoT-based Healthcare System for Smart Hospital with advanced technology is analyzed in this section. The data analysis is a vital role to enhance the high level performance of the developed system.

In that IoT-based RHS, Arduino Uno board, human body temperature sensors, heart rate sensor and oxygen saturation level sensor, four relay modules and SIM900 GSM module are important for designing the embedded device for sensing and storing the data in to cloud.

An embedded system designed for health care monitoring system by using three body sensors and important components is shown in Figure 11.3.

There are dissimilar BPM, SPO2, and temperature values for five diverse activities of human as the result exposed in Figure 11.5 of Multivariate Analysis. The performance of the developed system meets the 99% accuracy for practical usage in hospitals.

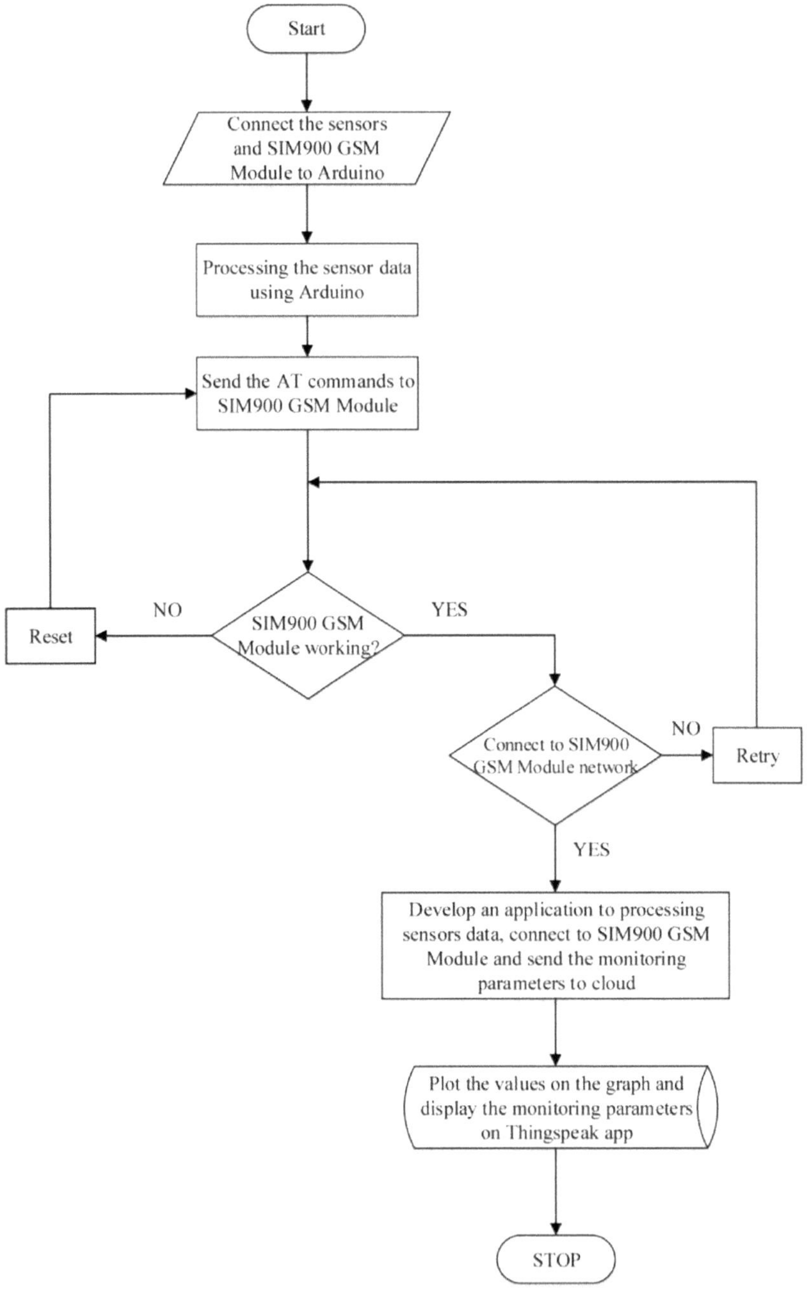

FIGURE 11.2 Developed IoT-based healthcare system.

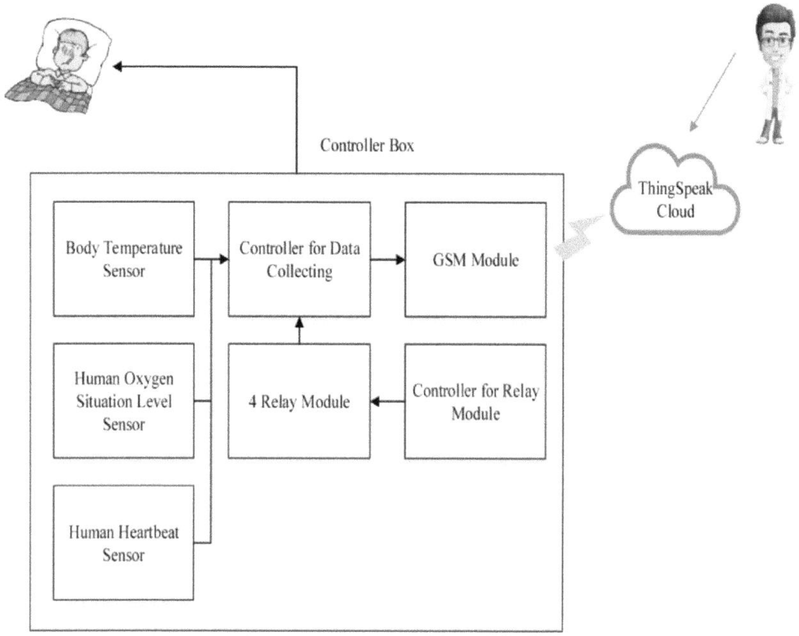

FIGURE 11.3 Embedded system and design.

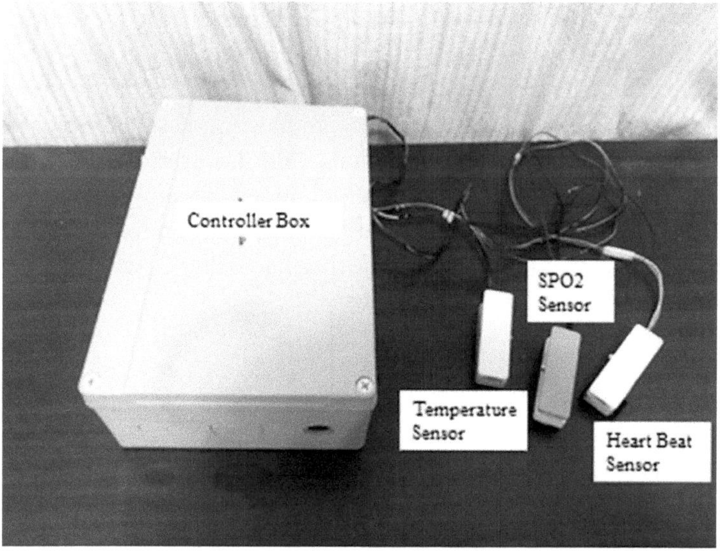

FIGURE 11.4 Developed system for IoT-based Healthcare design for smart hospital.

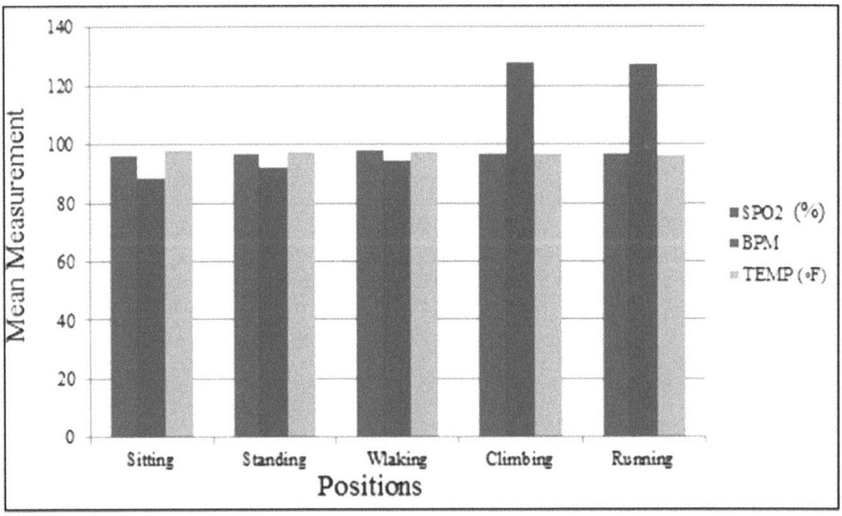

FIGURE 11.5 Experimental results.

11.8 DISCUSSION

The blockchain technology is very advanced in healthcare system and smart systems for state-of-the-art applications in modern society. The smart IoT-based embedded system development plays a crucial role to enhance the smart city implementation. The combination of blockchain technology and IoT-based system design has impressive efficient impacts on the applications at smart hospitals. The accuracy percentage of the measurement test could be found in high level than any other systems that were developed in recent days. The recommendation for IoT-based smart healthcare system with blockchain technology is high performance system for smart hospital design of today's.

11.9 CONCLUSION

Regular gadgets have become independent and intelligent in the IoT innovation. In any case, in the security space (information dependability), there are a few difficulties. There is a critical need to convey trust in tremendous approaching data sources like this. It is vital to have the option to forestall and distinguish existing dangers and the capacity to figure out common threats and assaults from now on. Consequently, we contend that there is a requirement for a more profound examination of prophetic IoT security. Over the last ten years, the ascent of EHRs and clinical gadgets has changed the medical services industry by modifying instructions to catch and oversee well-being data. In light of our participatory plan overview, the data from an EHR are not sufficient to make an ideal clinical choice. Most of our evaluators have brought up, for example, that there is no powerful technique for deciding or checking how a patient's well-being has developed within a couple of months before a clinical visit. All things being equal, most doctors choose to find or set up treatment plans in light of a preview of suggestive data caught during the patient experience or the patient's emotional depiction, which presents a need to get to more granular and ongoing patient well-being data.

KEYWORDS

- **IoT**
- **blockchain**
- **health record**
- **RHS**

REFERENCES

1. Tawalbeh, L.; Muheidat, F.; Tawalbeh, M.; Quwaider, M. IoT Privacy and Security: Challenges and Solutions. *Appl. Sci.* **2020,** *10*, 4102.
2. Srinivasu, P.; SivaSai, J.; Ijaz, M.; Bhoi, A.; Kim, W.; Kang, J. Classification of Skin Disease Using Deep Learning Neural Network with Mobilenet V2 and LSTM. *Sensors* **2021,** *21*, 2852.
3. NagaSrinivasu, P.; Rao, C. S. A Multilevel Image Encryption Based on Duffifing Map and Modified DNA Hybridization for Transfer Over an Unsecured Channel. *Int. J. Comput. Appl.* **2015,** *120*, 1–4.
4. Jiang, S.; Cao, J.; Wu, H.; Yang, Y. Fairness-Based Packing of Industrial IoT Data in Permission Blockchains. *IEEE Trans. Ind. Informatics* **2020,** *PP*, 1.
5. Wu, H.; Cao, J.; Yang, Y.; Tung, C.L.; Jiang, S.; Tang, B.; Liu, Y.; Wang, X.; Deng, Y. Data Management in the Supply Chain Using Blockchain: Challenges and a Case Study. In: *Proceedings of the 2019 28th International Conference on Computer*

Communication and Networks (ICCCN), Valencia, Spain, 29 July–1 August 2019; pp 1–8.
6. Dwivedi, A. D.; Srivastava, G.; Dhar, S.; Singh, R. A Decentralized Privacy-Preserving Healthcare Blockchain for IoT. *Sensors* **2019,** *19*, 326.
7. El-Rashidy, N.; El-Sappagh, S.; Islam, S.M.R.; M El-Bakry, H.; Abdelrazek, S. Mobile Health in Remote Patient Monitoring for Chronic Diseases: Principles, Trends, and Challenges. *Diagnostics* **2021,** *11*, 607.
8. Alam, T.;. Internet of Things and Blockchain-Based Framework for Coronavirus (COVID19) Disease. SSRN Electronic Journal, **2020**.
9. Lee, S.; Sim, K.;. Design and Hardware Implementation of a Simplified DAG-Based Blockchain and New AES-CBC Algorithm for IoT Security. *Electronics* **2021,** *10* (9), 1127.
10. Wang, D.; Wang, H.; Fu, Y. Blockchain-Based IoT Device Identification and Management in 5G smart grid. *EURASIP J. Wireless Commun. Network.* **2021,** *2021* (1).
11. Deloitte Switzerland. Can Blockchain Accelerate Internet of Things (IoT) Adoption|Deloitte Switzerland? (accessed 8 Sept 2021).
12. Atlam, H.; Azad, M.; Alzahrani, A.; Wills, G. A Review of Blockchain in Internet of Things and AI. *Big Data Cogn. Comput.* **2020,** *4* (4), 28.
13. Panigrahi, R.; Borah, S.; Bhoi, A.; Ijaz, M.; Pramanik, M.; Jhaveri, R.; Chowdhary, C. Performance Assessment of Supervised Classifiers for Designing Intrusion Detection Systems: A Comprehensive Review and Recommendations for Future Research. *Mathematics* **2021,** *9*, 690.
14. Pajooh, H.; Rashid, M.; Alam, F.; Demidenko, S. Multi-Layer Blockchain-Based Security Architecture for the Internet of Things. *Sensors* **2021,** *21*, 772.
15. Saura, J.R.; Ribeiro-Soriano, D.; Palacios-Marqués, D. Setting Privacy "by Default" in Social IoT: Theorizing the Challenges and 587 Directions in Big Data Research. *Big Data Res.* **2021,** *25*, 100245. https://doi.org/10.1016/j.bdr.2021.100245.
16. Rahman, Z.; Yi, X.; Khalil, I. Blockchain-Based AI-Enabled Industry 4.0 CPS Protection Against Advanced Persistent Threat. *IEEE Internet of Things J.* **2022,** 1–1. DOI: 10.1109/JIOT.2022.3147186.
17. Pradhan, D.; Sahu, P. K.; Goje, N. S.; Ghonge, M. M.; Tun, H. M.; Rajeswari, R.; Pramanik, S. Security, Privacy, Risk, and Safety Toward 5G Green Network (5G-GN). *Cyber Secur. Netw. Secur.* **2022,** 193–216.
18. Pradhan, D.; Sahu, P. K.; Ghonge, M. M.; Tun, H. M. Security Approaches to SDN-Based Ad hoc Wireless Network Toward 5G Communication. In: *Software Defined Networking for Ad Hoc Networks*; Springer: Cham, **2022**; pp 141–156.
19. Pradhan, D. 5G-Green Wireless Network for Communication with Efficient Utilization of Power and Cognitiveness. In: *International Conference on Mobile Computing and Sustainable Informatics*; Springer: Cham, **2020**; pp 325–335.
20. Ghonge, M. M.; Mane, S.; Pradhan, D. Demystifying the Role of Blockchain Technology in Healthcare and Transaction. In: *Blockchain Technologies and Applications for Digital Governance*; IGI Global, **2022**; pp 60–84.
21. Thaung, S. M.; Tun, H. M.; Win, K. K. K.; Than, M. M. Exploratory Data Analysis Based on Remote Health Care Monitoring System by Using IoT. *Communications* **2020,** *8* (1), 1–8. DOI: 10.11648/j.com.20200801.11

22. Pradhan, D.; Tun, H. M.; Dash, A. K. IoT: Security & Challenges of 5G Network in Smart Cities. *Asian J. Convergence Technol.* **2022,** *8* (2), 45–50.
23. Ghonge, M. M.; Mane, S.; Pradhan, D. (2022). Demystifying the Role of Blockchain Technology in Healthcare and Transaction. In: *Blockchain Technologies and Applications for Digital Governance*; IGI Global, **2022**; pp 60–84.

CHAPTER 12

Security and Privacy of EHRs Sharing Through Blockchain Technology in Smart Cities

NITIN S. GOJE[1], DEBASIS DAS[1], N. R. ASHA RANI[2], SUNIL KUMAR BEHERA[3], and DEVASIS PRADHAN[4]

[1]*Department of MIS, Webster University, Tashkent, Uzbekistan*

[2]*Department of Civil Engineering, Alliance College of Engineering and Design, Alliance University, Bangalore, India*

[3]*I & C Engineering, Reliance Jio, Odisha, India*

[4]*Department of Electronics & Communication Engineering, Acharya Institute of Technology, Bangalore, India*

ABSTRACT

Healthcare information is exceptionally delicate as numerous medical care associations will be extremely hesitant to share health information. In any case, sharing the medical services information has a lot more applications for both the patients just as the examination establishments as well. Additionally, the current Electronic Healthcare Record (EHR) board framework will be put away in the focal data set in the type of plain text. The fast improvement of blockchain innovation advances a protected medical care framework, including clinical records just as persistent related information. This innovation gives patients broad, unaltered records and gives admittance to EHRs liberated from specialist organizations and treatment sites. The protection of the equivalent from digital attacks alongside protection

5G Green Communication Networks for Smart Cities. Devasis Pradhan, Rajeswari, Hla Myo Tun, Naw Khu Say Wah, & Thandar Oo (Eds.)

conservation through verified admittance is one of the huge difficulties for the healthcare area. For this reason, the utilization of blockchain- based networks can prompt a significant decrease in the weaknesses of the medical services frameworks and secure their information.

12.1 INTRODUCTION

Blockchain innovation is decentralized, alters apparent, and tempers safe innovation. Blockchain innovation depends on the idea of appropriated record innovation where each hub associated with the organization keeps a duplicate of the record. Blockchain innovation came into light in the year 2008 when satoshinakamo gave the idea of digital forms of money and the thought was executed in 2009. Since then blockchain innovation has advanced many folds and as a result of many benefits, the innovation is being utilized in numerous areas. Numerous associations are getting advantages of safety and security with the execution of blockchain innovation. Electronic clinical records are the records of an individual, which can be put away electronically on any gadget. The openness of the health records should be possible continuously from any place and stay with the individual all over. Be that as it may, with the progression of electronic clinical data, a few sorts of risk and responsibility are additionally implied where the security and corrupted details are the most serious matter implied with HER.

TABLE 12.1 Discuss About the Data Security and Privacy with Block Chain Technology.

Key Features for Block Chain Technology Inclined to EHR	
Security	• It ensures clinical information is on the way and very still, with the goal that information classification, uprightness, and accessibility can be kept up with.
	• Cryptography- information encryption, computerized mark, and access control instruments can guarantee secure information access in a solitary space.
Privacy	• Protection is an intently related idea to security yet has its own focuses, that is, it guarantees that individual information are gathered, utilized, secured, and annihilated lawfully and reasonably.

12.2 IMPORTANCE OF EHRS IN SMART CITIES

EHRs are an important component of smart cities. Smart cities leverage technology and data to improve the quality of life for citizens and EHRs play a crucial role in achieving this objective. Here are some reasons why EHRs are important in smart cities:

a) Improved Health Outcomes: EHRs can help healthcare providers access accurate and up-to-date information about a patient's medical history, medications, and allergies. This information can help providers make informed decisions about patient care and can improve health outcomes.
b) Better Coordination of Care: EHRs can facilitate better communication and coordination of care between healthcare providers, patients, and caregivers. In a smart city, EHRs can be used to connect patients with healthcare providers, schedule appointments, and provide access to telemedicine services.
c) Enhanced Population Health Management: EHRs can provide valuable insights into population health trends, allowing healthcare providers and public health officials to identify health risks and develop targeted interventions to improve health outcomes.
d) Increased Efficiency: EHRs can streamline administrative tasks, reduce errors, and save time and money for healthcare providers. In a smart city, EHRs can be integrated with other technology systems, such as mobile health apps and wearable devices, to provide a more comprehensive view of a patient's health status.
e) Better Data Analytics: EHRs can generate large amounts of data that can be used for research and data analytics. In a smart city, this data can be analyzed to identify trends and patterns in health outcomes, which can inform policy decisions and resource allocation.

In summary, EHRs are a critical component of smart cities. They can improve health outcomes, enhance population health management, increase efficiency, and provide valuable data for research and data analytics. They help in better clinical decision making by integrating patient information from multiple sources.

12.3 E-HEALTH RECORD

EHR is a digital version of a patient's medical record that is stored electronically and can be accessed by authorized healthcare providers from various locations. EHRs typically contain information such as a patient's medical history, diagnoses, medications, lab results, and other relevant health information. EHRs have several advantages over traditional paper-based medical records. They allow for easy access to patient information, reducing the time and effort required to find and organize paper records. EHRs also enable healthcare providers to share patient information easily and securely, which can improve coordination of care and reduce errors.

Another benefit of EHRs is that they can help healthcare providers make better-informed decisions about patient care. EHRs provide a complete and accurate view of a patient's medical history, allowing providers to quickly identify potential health issues, allergies, and drug interactions. EHRs also offer several benefits to patients. Patients can access their own health records online, allowing them to better understand their health status and take an active role in managing their own care. Patients can also share their health information with other healthcare providers, reducing the need for redundant tests and procedures. EHRs are an important tool in modern healthcare. They improve access to patient information, facilitate secure sharing of information between healthcare providers, and enable better-informed decisions about patient care. EHRs also empower patients to take a more active role in managing their own health.

12.3.1 BENEFITS OF E-HR

EHR has various advantages for the medical services environment:

a) Information accessibility in an organized structure brings down managerial expenses and odds of clinical mistakes.
b) A combined archive of a singular's health data improves information mining and investigation abilities.
c) Better openness to health records encourages patient responsibility for and wellness
d) Technology-supported frameworks enhance working and managerial costs.
e) Transparent data accessibility further develops doctor effectiveness and nature of care.
f) Data-driven independent direction further develops esteem-based repayment eligibility.

12.3.2 CHALLENGES IN ENABLING HER

The shortfall of interoperability is bringing about colossal misfortunes (financial and in any case); notwithstanding, the quantity of drives the medical care environment has started to address the circumstances have been far underneath assumption.[24,25] Many variables add to partner hesitance as to EHR interoperability:

a) **Misaligned inducement:** The partners' motivators are skewed concerning the objectives of interoperability.
   i. **Suppliers:** Every EHR merchant as of now sells its own exclusive frameworks. On the off chance that interoperability becomes a reality, their clients will at this point do not experience exclusive lock-in and will be allowed to pick any merchant.
   ii. **Buyers:** Current EHR framework clients have contributed billions of dollars in purchasing and setting up heritage EHR frameworks.

b) **Information security:** The vision of a solitary interoperable stage that possibly makes all medical care information available in one area raises significant security and security concerns.

c) **Absence of information sharing:** Traditionally, information dividing among payers and suppliers has been an issue, chiefly in light of the fact that they were working freely and did not share any normal information trade norms.

12.4 BLOCKCHAIN TECHNOLOGY

Blockchain technology, advocated by digital currencies like Bitcoin, stands to hold the keys to a considerably more far reaching and secure arrangement of clinical records. This is conceivable due to what a blockchain innately is: a record cryptographically upheld to guarantee the uprightness of the information on it. At the point when given something to do to get clinical information, blockchain can store data in a way that is open to anybody in the organization, totally permanent, and sealed. Basically, this technology resembles a connected rundown sort of information structure wherein hubs are associated with a pointer. The subsequent block consists of newly generated hash value with previous block details. The blocks have certain information such as:

a) The principal block made and added into the blockchain is called a Genesis block (GB) and this block does not have any previous hash esteem.
b) Every node or client can make a block of exchange information or transaction and send it to the blockchain network. The clients of the blockchain network should go for an agreement to add this block in the chain of blocks relying on the agreement calculation conveyed on the chain.

12.4.1 PROCESS OF BLOCKCHAIN

Generally, blockchain process depends on the transactions of blocks or data that need to be checked and legitimate. A blockchain network comprises different nodes associated in a distributed and decentralized manner. Any node that necessities to play out an transaction needs to compose the transaction in a block. A block can be considered as a compartment-like information construction that can comprise around 500 transaction with the size of a block to be 1MB. Figure 12.1 depicts the process of blockchain.

12.4.2 PRIVACY AND SECURITY

Blockchain technology can help improve the privacy and security of EHRs by providing a tamper-resistant and secure platform for storing and sharing patient information. One way that blockchain can enhance privacy and security is through the use of cryptographic techniques. Patient information can be encrypted and stored on the blockchain in a way that only authorized parties can access it. This can help prevent unauthorized access to patient records and protect patient privacy. Another way that blockchain can improve the security of EHRs is through its decentralized nature. Instead of relying on a central authority to manage patient records, blockchain allows for a distributed network of nodes to store and verify data. This can help prevent a single point of failure and reduce the risk of data breaches or cyber-attacks. Furthermore, blockchain technology can enable patients to have greater control over their own health data. Patients can use blockchain-based platforms to grant permission for healthcare providers to access their records and specify the types of data that can be accessed. This can help improve patient privacy and give patients greater control over their own health information. In addition to the above statements, blockchain technology has the potential to enhance the privacy and security of EHRs. By using cryptographic techniques and decentralization, blockchain can help prevent unauthorized access to patient records and reduce the risk of data breaches. Additionally, blockchain-based platforms can give patients greater control over their own health information, which can improve patient privacy and empower patients to take a more active role in managing their own care.

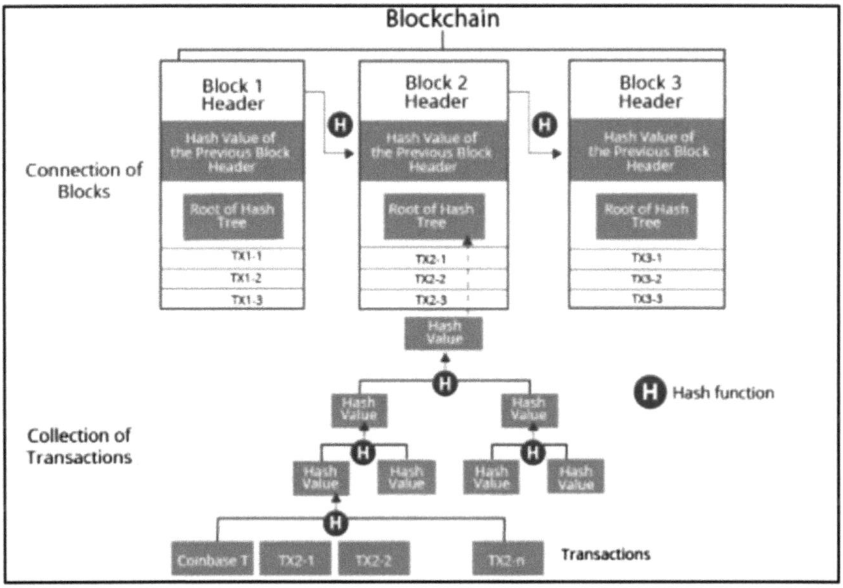

FIGURE 12.1 Blockchain process.

Source: Reprinted from Intellipat. https://intellipaat.com/blog/tutorial/blockchain-tutorial/how-does-blockchain-work/

12.5 HEALTHCARE DATA SYSTEM (HDS)

HDS are information systems designed specifically for managing and analyzing healthcare data. These systems play a crucial role in modern healthcare, as they allow healthcare providers to collect, store, and analyze patient information to improve patient outcomes and inform decision making. HDS can be used for a wide range of healthcare applications, including patient management, clinical decision support, quality improvement, and research. These systems typically consist of software applications, databases, and analytical tools that work together to manage healthcare data and generate insights.

One important application of HDS is EHRs. EHRs are digital records of patient health information that can be stored and accessed by healthcare providers from various locations. HDS can be used to create and manage EHRs, allowing healthcare providers to access patient information quickly and easily. HDS can also be used for clinical decision support. These systems can analyze patient data in real time and provide healthcare providers with recommendations for diagnosis and treatment. This can help healthcare providers make more informed decisions and improve patient outcomes. Additionally, HDS can be used for quality improvement initiatives. These systems can be used to analyze patient data and identify areas where improvements can be made in patient care. HDS can also be used to track performance metrics and monitor progress over time.

12.5.1 SERVICES

Since the healthcare data goes past the clinical reports and the individual healthcare data, the connected administrations additionally go past analysis by specialists and clinical trials. The extent of these administrations is different. They range from individual patients through conclusion to hierarchical and public level by means of medical care asset sending. Aside from this, the working of the administrations identified with medical care research programs, drug circulation, and health care coverage systems require assorted medical care information as fundamental information.

12.5.2 SECURITY IN STORAGE SYSTEM

The storage-driven security issues emerge principally because of unified stockpiling and the board by medical care-related associations and are additionally arranged into weaknesses and dangers. A critical security impediment regarding unified medical care information

TABLE 12.2 Features of HDS.

Sl. no.	Features	Remarks
1	EMR	• It contains the patient's medical history, diagnoses, medications, lab results, and other relevant health information.
2	EHR	• a digital record of a patient's health information. • containing a patient's medical history, diagnoses, medications, allergies, lab results, and other relevant health information. • an important tool in modern healthcare, helping to improve patient outcomes and coordination of care.
3	MPI	• It maintains accurate and complete patient identification and demographic information, which is essential for effective patient care and coordination. • It is used to link patient data from various sources, such as EHRs, laboratory test results, imaging studies, and other healthcare systems.
4	PP	• The portal allows patients to view their medical records, including test results, medications, immunizations, and allergies, and to communicate with their healthcare providers. • Patient portals are designed to improve patient engagement and to enable patients to take a more active role in their healthcare.
5	RPM	• Healthcare providers to monitor patients' health status and receive data from patients outside of traditional healthcare settings • RPM can also improve patient engagement by enabling patients to take a more active role in their healthcare. • RPM can also reduce the need for in-person visits, making healthcare more convenient and accessible for patients.
6	CDS	• It provides healthcare providers with actionable information and guidance at the point of care • CDS systems can take many forms, including alerts, reminders, order sets, guidelines, and predictive models. • CDS systems can help healthcare providers to improve patient safety, reduce medical errors, and enhance the quality of care.
7	PMS	• It helps healthcare providers manage their practice operations, including patient scheduling, billing, and administrative tasks. • PMS systems can automate many routine administrative tasks, freeing up staff time and improving efficiency.

EMR, Electronics Medical Report; EHR, Electronics Health Record; MPI, Master Patient Index; PP, patient portals; RPM, Remote Patient Monitoring; CDS, Clinical Decision Support; PMS Practice Management Software.

stockpiling frameworks is the weakness because of a solitary mark of control.

12.5.3 *BLOCKCHAIN IMPROVES DATA SECURITY AND PROTECTION*

Subsequent Blocks are connected through hash upsides of square information. When the information gets recorded on the record, it is difficult to adjust it because of its permanence property. It forestalls unapproved change just as accommodates the detectability of medical care information. There is no brought-together capacity of the blockchain information. All things being equal, each partaking hub in the decentralized organization keeps a duplicate of the blockchain put away with it.

12.6 LIMITATION OF BLOCKCHAIN TECHNOLOGY

A major test for medical care information frameworks is the means by which to accumulate, store, and break down close to home medical services information without raising security infringement. For such frameworks, protection concerns have demonstrated hindrances to take on and the absence of sufficient security. The blockchain is viewed as a solid stage since all activities made by framework members are recorded on the chain, however, the extending chain makes it computationally testing to change any block without location.

12.7 CONCLUSION AND FUTURE DIRECTIONS FOR BLOCKCHAIN

As blockchain innovation is arising, its application in the medical services area is additionally under experimentation. Subsequently, a considerable lot of the chose works were restricted to model-based approvals. Review on interoperability in EHR has been restricted to the expressed three regions, for example, structure and semantic interoperability, patient-driven security conservation, cross-chain interoperability. The deals with security protecting methods are restricted to the works in medical services information security, and the assessment is finished in view of restricted execution and consistency measurements.

KEYWORDS

- **EHRs**
- **blockchain**
- **data privacy**
- **decentralized technology**
- **attribute-based encryption**
- **data sharing**
- **security**
- **integrity**

REFERENCES

1. Persons, K. R.; Nagels, J.; Carr, C.; Mendelson, D. S.; Primo, H. R.; Fischer, B.; Doyle, M. Interoperability and Considerations for Standards-Based Exchange of Medical Images: HIMSS-SIIM Collaborative White Paper. *J. Digit. Imaging* **2020,** *33*, 6–16.
2. Sorace, J.; Wong, H. H.; DeLeire, T.; Xu, D.; Handler, S.; Garcia, B.; MaCurdy, T. Quantifying the Competitiveness of the Electronic Health Record Market and Its Implications for Interoperability. *Int. J. Med. Inf.* **2020,** *136*, 104037.
3. Watford, S.; Edwards, S.; Angrish, M.; Judson, R. S.; Friedman, K. P. Progress in Data Interoperability to Support Computational Toxicology and Chemical Safety Evaluation. *Toxicol. Appl. Pharmacol.* **2019,** *380*, 114707.
4. Dubovitskaya, A.; Baig, F.; Xu, Z.; Shukla, R.; Zambani, P. S.; Swaminathan, A.; et al. ACTION-EHR: Patient-Centric Blockchain-Based Electronic Health Record Data Management for Cancer Care. *J. Med. Internet Re.* **2020,** *22* (8), e13598.
5. Hathaliya, J. J.; Tanwar, S. An Exhaustive Survey on Security and Privacy Issues in Healthcare 4.0. *Comput. Commun.* **2020,** *153*, 311–335.
6. Zhuang, Y.; Sheets, L. R.; Chen, Y. W.; Shae, Z. Y.; Tsai, J. J.; Shyu, C. R. A Patient-Centric Health Information Exchange Framework Using Blockchain Technology. *IEEE J. Biomed. Health Inf.* **2020,** *24* (8), 2169–2176.
7. Tanwar, S.; Parekh, K.; Evans, R. Blockchain-Based Electronic Healthcare Record System for Healthcare 4.0 Applications. *J. Inf. Secur. App.* **2020,** *50*, 102407.
8. Belchior, R.; Vasconcelos, A.; Guerreiro, S.; Correia, M. A Survey on Blockchain Interoperability: Past, Present, and Future Trends. *ACM Comput. Surv. (CSUR)* **2021,** *54* (8), 1–41.
9. Al Omar, A.; Bhuiyan, M. Z. A.; Basu, A.; Kiyomoto, S.; Rahman, M. S. Privacy-Friendly Platform for Healthcare Data in Cloud Based on Blockchain Environment. *Fut. Gen. Comput. Syst.* **2019,** *95*, 511–521.

10. Rajput, A. R.; Li, Q.; Ahvanooey, M. T. A Blockchain-Based Secret-Data Sharing Framework for Personal Health Records in Emergency Condition. *Healthcare* **2021,** *9* (2), 206.
11. Garrido, A.; Lopez, L. J. R.; Álvarez, N. B. A Simulation-Based AHP Approach to Analyze the Scalability of EHR Systems Using Blockchain Technology in Healthcare Institutions. *Inf. Med. Unlocked* **2021,** *24*, 100576.
12. Chenthara, S.; Ahmed, K.; Wang, H.; Whittaker, F.; Chen, Z. Healthchain: A Novel Framework on Privacy Preservation of Electronic Health Records Using Blockchain Technology. *Plos One* **2020,** *15* (12), e0243043.
13. Shuaib, M.; Daud, S. M.; Alam, S.; Khan, W. Z. Blockchain-Based Framework for Secure and Reliable Land Registry System. *TELKOMNIKA (Telecommun. Comput. Electron. Control)* **2020,** *18* (5), 2560–2571.
14. Shuaib, M.; Daud, S. M.; Alam, S. Self-Sovereign Identity Framework Development in Compliance with Self Sovereign Identity Principles Using Components. *Int. J. Mod. Agric.* **2021,** *10* (2), 2021.
15. Raghuvanshi, A.; Singh, U. K.; Shuaib, M.; Alam, S. An Investigation of Various Applications and Related Security Challenges of Internet of Things. *Mater. Today: Proc.* **2021**. DOI: 10.1016/j.matpr.2021.03.096.
16. Akkiraju, R.; Sinha, V.; Xu, A.; Mahmud, J.; Gundecha, P.; Liu, Z. et al. Characterizing Machine Learning Processes: A Maturity Framework. In: *Business Process Management: 18th International Conference, BPM* **2020**, Seville, Spain, 13–18 Sept 2020, *Proceedings 18*; Springer International Publishing, 2020; pp 17-31.
17. Qu, W.; Wu, L.; Wang, W.; Liu, Z.; Wang, H. A Electronic Voting Protocol Based on Blockchain and Homomorphic Signcryption. *Concurrency Comput.: Practice Exp.* **2022,** *34* (16), e5817.
18. Pradhan, D.; Tun, H. M.; Dash, A. K. IoT: Security & Challenges of 5G Network in Smart Cities. *Asian J. Convergence Technol. (AJCT)* **2022,** *8* (2), 45–50.
19. Tanwar, S.; Parekh, K.; Evans, R. Blockchain-Based Electronic Healthcare Record System for Healthcare 4.0 Applications. *J. Inf. Secur. App.* **2020,** *50*, 102407.
20. Usman, M.; Qamar, U. Secure Electronic Medical Records Storage and Sharing Using Blockchain Technology. *Procedia Comput. Sci.* **2020,** *174*, 321–327.
21. Wah, N. K. S. Integration of AI/Ml in 5G Technology toward Intelligent Connectivity, Security, and Challenges. *Mach. Learn. Algorithms App. Eng.* **2023,** *239*.
22. Shamshad, S.; Mahmood, K.; Kumari, S.; Chen, C. M. A Secure Blockchain-Based E-Health Records Storage and Sharing Scheme. *J. Inf. Secur. App.* **2020,** *55*, 102590.
23. Bodkhe, U.; Tanwar, S.; Parekh, K.; Khanpara, P.; Tyagi, S.; Kumar, N.; Alazab, M. Blockchain for Industry 4.0: A Comprehensive Review. *IEEE Access* **2020,** *8*, 79764–79800.
24. Abu-Elezz, I.; Hassan, A.; Nazeemudeen, A.; Househ, M.; Abd-Alrazaq, A. The Benefits and Threats of Blockchain Technology in Healthcare: A Scoping Review. *Int. J. Med. Inf.* **2020,** *142*, 104246.
25. Cilliers, L. Wearable Devices in Healthcare: Privacy and Information Security Issues. *Health Inf. Manag. J.* **2020,** *49* (2–3), 150–156.
26. Priyanka, K.; Mallavaram, G.; Raj, A.; Pradhan, D. Cognitiveness of 5G Technology Toward Sustainable Development of Smart Cities. *Decis. Support Syst. Smart City App.* **2022,** 189–203.

27. Pradhan, D.; Tun, H. M. Security Challenges: M2M Communication in IoT. *J. Electr. Eng. Autom.* **2022,** *4* (3), 187–199.
28. Pharma, K. D. S. D. *Exscientia Achieve Breakthrough in AI Drug Discovery: Healthcare IT News-Portland*; Healthcare IT News, ME, 2020.
29. Langner, S.; Häse, F.; Perea, J. D.; Stubhan, T.; Hauch, J.; Roch, L. M. et al. Beyond Ternary OPV: High-Throughput Experimentation and Self-Driving Laboratories Optimize Multicomponent Systems. *Adv. Mater.* **2020,** *32* (14), 1907801.
30. Zhang, X.; Xiao, C.; Glass, L. M.; Sun, J. DeepEnroll: Patient-Trial Matching with Deep Embedding and Entailment Prediction. In: *Proceedings of The Web Conference*; **2020**; pp 1029–1037.
31. Ghonge, M. M.; Mane, S.; Pradhan, D. Demystifying the Role of Blockchain Technology in Healthcare and Transaction. In: *Blockchain Technologies and Applications for Digital Governance*; IGI Global, **2022**; pp 60–84.
32. Pradhan, D.; Dash, A.; Tun, H. M.; Wah, N. K. S.; Oo, T. Improvement of Capacity and QoE: Distributed Massive MIMO (DM-MIMO) Technology-5G. *J. Netw. Secur. Comput. Netw.* **2022,** *8* (3), 9–17. https://doi.org/10.46610/jonscn.2022.v08i03.002
33. Pradhan, D.; Dash, A.; Tun, H. M.; Wah, N. K. S.; Oo, T. A Sustainable Key Enabler for mm-Wave Beamforming in 5G Environment. *J. VLSI Design Signal Process.* **2022,** *8* (3), 10–17. https://doi.org/10.46610/jovdsp.2022.v08i03.002
34. Pradhan, D.; Tun, H. M.; Naing, Z. M.; OO, T. Green WPC: Energy Harvesting in Smart Cities. *Middle East Res J. Eng. Technol.* **2022,** *2* (1), 8–12.

CHAPTER 13

Bigdata Analytics for Supply Chain Management Toward 5G Green Cities

SUBHASH CHANDER VERMA[1], HLA MYO TUN[2], TARIQUE AKHTAR[3], and ALESSANDRO BRUNO[4]

[1]*Data Scientist, Wipro Limited, Bengaluru, India*

[2]*Research Department, Yangon Technological University, Yangon, Myanmar*

[3]*Senior Data Scientist, Agility*

[4]*Doctor of Computer Engineering AI Scientist at IULM Libera Università di Lingue e Comunicazione di Milano, Milan, Italy*

ABSTRACT

Bigdata played an important role in international transportation; in particular, supply chain management (SCM) has the potential to have a significant impact on the industry as a whole. Future cities' transportation capacity is directly impacted by big data. Over the past 10 years, urbanization has significantly increased, and by 2050, one in three people will live in a city. Big data analytics is likely to be used to update the transportation infrastructure, which is necessary to keep up with the current flow of goods while also limiting its impact on the environment and human health. Smart cities are becoming increasingly popular as a solution to this issue.

5G Green Communication Networks for Smart Cities. Devasis Pradhan, Rajeswari, Hla Myo Tun, Naw Khu Say Wah, & Thandar Oo (Eds.)

13.1 INTRODUCTION

A smart city that makes use of ICT to enhance citizens' quality of life and the effectiveness of urban services is known as a smart city. The information generated by various city systems is combined to provide efficient services to accomplish this objective. This information comes from a lot of different places, like sensor devices that are installed in cars, buildings, streets, and other places. Smart cities are experiencing difficulties due to the heterogeneity of the stored data from various sources as a result of the increased number of installed IoT devices and sensors. For instance, Pourzolfaghar and Helfert (2017) identified this issue as one of the obstacles that prevent facility management companies from making use of the integration of information from various sources in buildings to provide useful services. The situation that big data analytics is going through is comparable to the development of smart cities.

The potential advantages and drawbacks of big data are significant issues for businesses in this day and age of industry-wide digitization. According to Shivarajah (2017), big data holds the promise of brand-new data-driven services that will enable novel products and business models as well as process enhancements. As a result, more and more businesses are putting their money into using big data. Cloud computing has made it possible to economically collect; a process that has led to a rise in the importance of BDA in supply chains. In addition, BDA for supply chains demonstrates significant potential for supply chain management (SCM) process enhancement, cost reduction, and improved decision-making.

13.2 BACKGROUND

13.2.1 SUPPLY CHAIN

The network of all the people, organizations, resources, activities, and technology involved in the production and sale of a product is called a supply chain. Everything from the supplier's delivery of raw materials to the manufacturer's delivery of finished goods to the end user is included in a supply chain. The distribution channel is a section of the supply chain that is involved in getting the finished product from the manufacturer to the consumer. In the meantime, supply chain managers are entitled to assert that they significantly contributed to the spread of the information technology revolution. As supply chain activities were integrated with the Internet, E-SCM, or e-supply chain management, was a significant transformation.[7] As a result, green or sustainable supply chains have emerged as an essential component of business and government operations in 5G Green Smart Cities.

13.2.2 BIG DATA

The process of collecting and storing large amounts of data for later analysis dates back a long time, even though the term "big data" is relatively new. Organizations obtain data from a variety of sources, including social media, machine-to-machine data, sensor data, and business transactions.[6] Data must be handled quickly because it moves at a rate never before seen. "Big Data" refers to data sets that general computers simply cannot adequately capture, manage, or process.

The technology that makes it simpler to store, manage, process, interpret, analyze, and visualize the enormous amount of data is now referred to by the name.[2] The initial technical definition of Big Data, which is referred to as the "3Vs"—velocity, variety, and volume—was provided in a research report that was published by Meta Group, which is now known as Gartner.[3] Later, the 7Vs in Figure 13.1 and the 5Vs, which stand for veracity, visualization, and value, were added to the definition. Big Data offers logistics companies a fresh source of competitive advantages in supply chain management. These benefits include increased visibility, the capacity to respond immediately to shifts in demand and capacity, insights into customer behavior and patterns that can be used to improve products and devise more cost-effective pricing, and the capacity to adapt to shifts in both.

13.2.3 BIG DATA ANALYTICS (BDAS)

BDA is a popular buzzword these days. It is not merely a fad. Instead, it is fundamentally altering every aspect of our lives, including transportation. Big data transformation typically involves two primary processes: Data management, followed by specific Data Analytics methods.[1,20] The investigation is characterized as "methods used to dissect and secure insight from enormous information," though the information the executives is characterized while storing information with suitable executable outcomes.

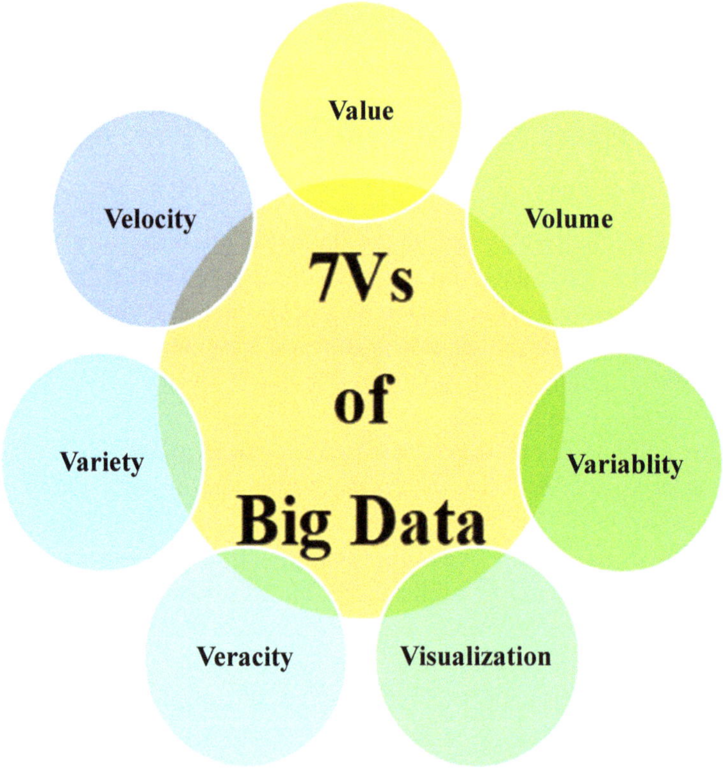

FIGURE 13.1 7Vs characteristics for big data.

13.2.4 BDA IN SUPPLY CHAIN

Members of the supply chain are linked together by a significant physical flow of whole architecture of the business model including all the basic components. Manufacturers have been looking into novel technologies and strategies in order to gain and maintain a competitive advantage in the globalized business environment of today and the ever-increasing competition. The significance of BDA in SCM is increasing. Since a lot of SCM-related research on BDA has been published over the past 10 years, many review articles have been written about it. From both a top-down and bottom-up perspective, our study examines the BDA in SCM research and addresses some of the flaws in the published literature while also adding to it.

13.2.5 5G

A novel networking solution known as 5G has been developed to meet the communication needs of the future. Numerous smart devices can now communicate with one another at any time and from anywhere thanks to this technology. These devices require extreme communication network capabilities, such as Gbps-scale data rates and delays close to zero. The 5G clears the way for a Factory for the Future (FoF) with these capabilities for Industry 4.0 and Supply Chain 4.0. The idea of having general-purpose computing and storage assets has been realized as a result of cloud computing (CC), software-defined networking (SDN), and network function virtualization (NFV), all of which involve the virtualization of network functions (NFVs), which enables scalability, adaptability, and interoperability. Moving mobile edge computing to 5G systems results in the physical separation of the planes, allowing the data plane to remain close to the user at the network edge and the upload of the centralized control plane to cloud servers.

13.3 IMPACT OF 5G ON SUPPLY CHAIN MANAGEMENT IN SMART CITIES

Responding quickly is becoming increasingly important in modern supply chain management. Just before the system achieved low costs and productivity, it added responsiveness to make it flexible enough to meet the needs of the e-commerce market. As a result of the fact that many telecommunications service providers have already begun the construction of their infrastructure, 5G will soon be implemented globally, and the entire world expects its implementation to begin this year. The big data process is made possible by 5G technology, which is 100 times faster than 4G technology. Unmanned vehicles and automated robotics can now operate because the supply chain can be automated to capture data from multiple points and process it more quickly.

13.3.1 LOGISTICS

In today's digitalization frenzy, where nearly all businesses are looking to move their services and operations online, the increased speed and capacity to handle more traffic have become extremely important. 5G technology is a thousand times faster than 4G technology. As a result, many businesses and their operations can be connected simultaneously through the implementation of 5G technology in supply chain management, reducing logistical difficulties and delays.

13.3.2 CONTINUOUS MONITORING OR SURVEILLANCE

The absence of an effective monitoring system to prevent product damage during transportation is currently one of the biggest challenges facing supply chains. With the assistance of hardware solutions like 5G tracking devices, which aid in the 24-h, 7-day-a-week monitoring of the products that are in transit, the implementation of 5G technology into supply chain management can effectively address this issue.

13.3.3 ACCESS TO MORE RESOURCES

Parts of the world that were previously difficult to connect to the rest of the world are now accessible thanks to the introduction of 5G technology. Access to previously inaccessible resources has been made easier with improved connectivity. By making workload distribution simpler, it has contributed to improved supply chain management.

13.3.4 WAREHOUSE MANAGEMENT

The storage of goods, particularly perishables, is another issue plaguing the supply chain industry in addition to poor management of in-transit goods and products. With the introduction of 5G technology in supply chain management, it will be simple to achieve better management of warehouses and storage facilities. Hardware solutions like 5G sensors frequently update and notify important information like temperature, humidity, and so on. As a result, managing transport operations becomes simple.

13.3.5 EASY INTEGRATION TO OTHER TECHNOLOGIES

New technologies are blockchain, robotics, and artificial intelligence, among others, Additionally, their widespread use has made it clear that the supply chain management sector will eventually adopt it as well. In point of fact, in some instances, particular supply chains have already begun the integration procedures. In order to ensure a smooth transition to these technologies, supply chain management must incorporate 5G technology. This is due to the fact that 5G services are the only ones capable of providing the high bandwidth and low latency features necessary for the digital infrastructure needed to bring about this integration and provide cost-effective automatic data-sharing systems.

13.4 SUSTAINABLE SUPPLY CHAIN MANAGEMENT (S-SCM)

The integration of social, economic, and environmental goals to meet current needs without jeopardizing those of future generations is sustainability. In addition, the strategic and open management of supply chain activities through the incorporation of sustainability (environmental, social, and economic) goals into all processes is known as sustainable supply chain management (S-SCM).[1,2] This is done in order to meet the needs of stakeholders. The triple bottom line (TBL) strategy aims to achieve not only economic but also social and environmental goals by incorporating sustainability goals into supply chain activities. The S-SCM idea has previously been the subject of research. Transparency, risk management, strategy, and culture were identified as the four dimensions of S-SCM in the four-factor model proposed in the References.[2–8] Through the strategy dimension, businesses are encouraged to incorporate sustainability into their corporate strategy. It is more successful for a business to incorporate sustainability into its strategy than to pursue it on its own, with benefits felt throughout the supply chain.[10–12]

Operational-level activities must be closely monitored toward accomplishments in order to monitor the dimension's implementation throughout the supply chain.[7–9] Big data analytics technologies can make monitoring more effective and efficient.[7–12] As a result, the strategy aspect of SSCM is likely to be supported by big data analytics. It is still difficult to create a sustainable culture and keep an eye on it throughout the supply chain.

13.5 SMART SUPPLY CHAIN FRAMEWORK FOR SMART CITIES

"Smart supply chain" is used a lot in the literature, but its definitions and ideas are often vague and inconsistent. Figure 13.2 depicts the smart framework for integrated SCM with BDA. The term is frequently used to describe the various stages of the concept:

a) Smart supply chain drivers: such as smart transportation and logistics intelligent warehouse
b) Digital aspects and data streams: analyzing historical data repositories and real-time data flows with various analytics models
c) Aspects of technology: A SSCM is created when SC incorporates cutting-edge technologies like IoT and BD.

Smart cities include features like inclusiveness and sustainability, and there are numerous new Internet technology interfaces.[19] The term has been around since the middle of the 1990s, but it really took off in 2010 and 2011 when more cities started competing for innovation and sustainability. SCM has a lot of opportunities in smart cities. They might, for instance, provide platforms for open data from a variety of sources, which is essential for mobility in the supply chain.[22,23] In order to enhance vehicle routing and transportation planning, intelligent traffic systems, for instance, can quickly forecast specific locations.[24,25]

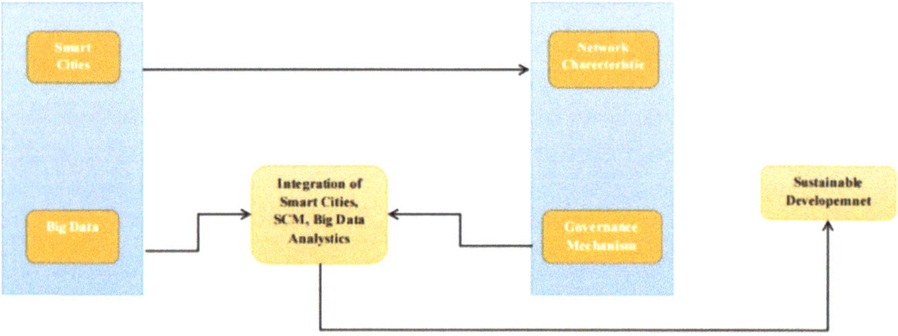

FIGURE 13.2 Smart framework for integrated smart cities with BDA and SCM.

The modern concept of a smart city is inextricably linked to new technology, urbanization, big data, sustainability issues, and other related topics. However, the term "smart cities" can refer to a wide variety of components, including energy-related solutions, comprehensive city planning, and other specialized infrastructures (Figure 13.2). Communities, local governments, businesses, or a combination of the two could take the initiative and integrate activities at the organizational and local levels.

13.6 APPLICATION OF BIG DATA IN SMART SCM

BDA is the application of cutting-edge analytical methods to massive data sets. It expands the dataset available for analysis beyond the typical ERP system internal data. In addition, it analyzes both new and old data sources with strong statistical tools. Fresh insights are generated as a result, assisting decision-makers in the supply chain with everything from strategic decisions regarding the best supply chain operating models to front-line operations.[19,20]

13.6.1 SC-NETWORK DESIGN (SC-ND)

The physical configuration of the supply chain, which has an impact on the majority of a company's business units or functional areas, must be determined for the supply chain network design project.

The following steps must be followed when designing a supply chain:

a) Establishment of long-term strategic goals;
b) Establishment of the scope of the designed project;
c) Selection of the analysis type to be executed for the designed product;
d) Identification of tools; and
e) Project completion.

13.6.2 SUPPLIER RELATIONSHIP MANAGEMENT

Sixty-four percent of supply chain executives believe that big data is an important and disruptive technology, according to the SCM World Report from September 2014. Big data enables businesses to enhance relationships and provide superior customer services. Businesses can also address distribution-related issues with the help of BDAs. BDA methodologies are used for precise data on the patterns of spending by an organization, which can be used to manage relationships with suppliers.[20–22]

13.6.3 DEMAND PLANNING

With the help of big data, a lot of executives in the supply chain want to improve demand forecasting and production planning.[25] In SCM, accurate demand forecasting has always been a significant challenge.[26] BDA can determine the optimal price data, demand signal, and trace consumer loyalty. However, the ability to implement cutting-edge hardware, software, and algorithm architecture is one of the obstacles that organizations must overcome.[27] BDA enables the identification of brand-new market trends and the underlying causes of problems, failures, and defects.

13.6.4 PRODUCT TRACEABILITY

It is essential to the operation of a successful supply chain. Utilizing barcode scanners and attaching radio frequency identification devices to specific products makes it simple for supply chain managers to track a product. With the help of big data analytics, businesses are able to collect precise product information so that managers can keep track of their distribution cycle. Food and beverage managers, for instance, will be able to predict when food spoilage is likely to occur with ease. The ability to track goods from production to retail is made possible by improved traceability. Businesses are able to streamline distribution through improved coordination with stakeholders in the supply chain thanks to enhanced traceability. Logistics firms would incur costs in the event of either an early or late delivery. One of the primary risks that logistics companies are concerned about is the gap in time between the expected and actual delivery. Supply chain management can benefit from using big data analytics to reduce the likelihood of delivery times that are incorrect. You will be able to improve traceability, which ensures that products can be traced from production to sale, with this assistance.[22–24]

13.6.5 CUSTOMIZE PRODUCTION

Manufacturers are able to use data analytics to thoroughly examine the entire supply chain conducting the verification process with all collective data. This ability can be utilized by manufacturers to identify inefficient processes, components, and bottlenecks. Even though the BDA of today has made it possible to accurately predict customer preferences and demands for customized products, centralized production, and large-scale production were irrational because they concentrated solely on the orders of a few selected customers. A few studies have examined the manufacturing application of BDA.[23]

13.6.6 MACHINE MAINTENANCE

The maintenance team's efficiency is being improved as a result of the growing use of big data in the field. The maintenance department's effectiveness is rising as a result of the shift toward data analytics, which is also increasing operational uptime. More than just replacing things before they break down, data analytics produce other outcomes.[24,25]

13.6.7 MANUFACTURING

Employee's performance metrics can be tracked, analyzed, and shared with the right application of BDA techniques. BDA methods are also used to find employees who are struggling or dissatisfied, as well as those who perform well or poorly. Instead of conducting annual studies based on human memory, these methods enable businesses for realistic monitoring of the whole process. Numerous electronic devices, digital machinery, and sensors are utilized in factories and production lines, generating a significant amount of data. As a result, intelligent factory shop floor logistics systems can be built with BDA.[24,29] Data analysis methods can be used to track defects, improve product quality, and enhance product manufacturing process activities in manufacturing.

13.6.8 FINANCE

Financial institutions aim to improve efficiency and sustainably increase their competitive advantage. These organizations must continually incorporate big data and appropriate analytical methods into their business strategies in order to maintain a sustainable competitive advantage and remain in business. Big data and analytical methods have improved significantly over the past few years, and a lot of money has been invested in them. Financial institutions and service providers can gain valuable knowledge and insights with the assistance of big data and analytical techniques.

13.6.9 HEALTHCARE INDUSTRY

It is essential to comprehend the driving forces and developments that will not only enhance the quality of care but also streamline efficiency within the hospital environment in order to advance from big data analytics. The way we think about and design efficient processes in the hospital will be significantly impacted by these developments. In a variety of ways, big data and telemedicine are working wonders in the medical and healthcare industries. Health informatics, which promises faster knowledge, analysis, and information sharing, has been established as a result of the enormous amount of health data generated by medical Internet-of-Things devices. Utilizing wearable devices, doctors can perform a diagnosis on the patient based on the health data they have collected, carry out the diagnosis, and enhance the patient's outcomes.

13.7 CHALLENGES OF BDA IN SCM

Businesses find it challenging to implement BDA because of growing concerns about data privacy and cyberattacks, despite the rapid growth of data.[56]

13.7.1 CHALLENGES FACED AT ORGANIZATIONAL LEVEL

The analytics process takes a long time because of both internal factors like a lack of access to data and external factors like the volume of Big Data, supply chain complexity, and interpretation goals for the datasets. If decision-makers react to insignificant changes in the physical world that would worsen the "bullwhip effect," supply chain risk and inventory cost can increase. There is an increased risk that decision-makers will identify statistically significant relations with insignificant causal linkage rather than irrelevant correlations due to large amount of data that is too precise.

13.7.2 TECHNICAL LEVEL

The performance of the analytics techniques' results can be affected by the quality of the stored and utilized data. Based on its sources and applications, data is intangible and multidimensional. The multidimensional datasets' dimensions can be categorized as intrinsic or contextual. The quality of the data should be consistent for results that are consistent and reliable for making decisions. The quality of the data that is collected in the supply chain may be affected by the variety of data and the types of data sources.

13.8 SECURITY AND PRIVACY

In supply chain management, data security is getting more and more important. Blockchain, IoT, and machine learning, among other emerging technologies, are increasing with a Supply Chain Network that collects data from various sources, analyzes it, and provides insights; data sharing is a major factor. However, a variety of privacy and security laws pertaining to data sharing may make it difficult for regional or global Supply Chain Networks to share data with various sources. In such situations, the accuracy of Big Data Analytics' potential insights may be compromised by a lack of shared data. In a similar vein, privacy is yet another aspect of BDA in the SCM space that requires additional investigation. As a result, ongoing research into the regulations and how they affect BDA is crucial.

13.9 IMPACT AND BENEFITS OF BDA IN S-SCM

Statistical models and integrated data from the entire supply chain can be used to improve demand forecasting. Planning for replenishment and inventory management are affected by this. For instance, if you have the opportunity to organize them ahead of time, you can be rest assured that nothing will ever run out. At this point, suitable data from the past and present but also data on macroeconomic factors, market trends, and even competition. We can see opportunities and intervene early on in problems by using supply chain analysis to analyze supplier's performance and compliance in real-time on-site quarterly or annually. In order for logistics businesses to implement an efficient data-driven business model, capitalizing data is an essential strategy. The application of big data in logistics can have a significant impact in three main areas: improved operational efficiency by increasing transparency, maximizing resource utilization, and enhancing performance and quality. A better experience for the customer in order to keep their trust and keep them coming back made use of the effective data-driven business model to generate more revenue. When businesses use the information they have gathered about their products to predict what their customers want, big data sets are useful.

13.10 CONCLUSION

Management of the supply chain has evolved into an essential part of how global economies work. BDA plays a crucial role in the operation of SCM through a variety of means, including the management of shifts in customer preference, enhancements in visibility, and resilience. Recent work in the present scenario has increased the uncertainties that exist in the supply chain. As a result, there has been a rise in research publications and practitioners' and academics' interest in BDA in SCM (e.g., Ref. [5]). To better comprehend BDA in SCM research, an interdisciplinary perspective is required, just as it was in the case of SCM research.[6] Due to its numerous effects on the organization and the world, sustainability in SCM is an important consideration. Prior research on sustainability in fifth G-enabled smart cities made use of theories, such as the stakeholder theory, according to our findings.

KEYWORDS

- **5G**
- **big data**
- **big data analytics**
- **supply chain**

REFERENCES

1. García-Holgado, A.; García-Peñalvo, F. J.; Butler, P. Technological Ecosystems in Citizen Science: A Framework to Involve Children and Young People. *Sustainability* **2020,** *12*, 1863.

2. Lierow, M. B2City: The Next Wave of Urban Logistics. *Supply Chain* **2014,** *247*, 41–48.
3. Burt, J. A. The Environmental Costs of Coastal Urbanization in the Arabian Gulf. *City* **2014,** *18*, 760–770.
4. Zambon, I.; Colantoni, A.; Carlucci, M.; Morrow, N.; Sateriano, A.; Salvati, L. Land Quality, Sustainable Development and Environmental Degradation in Agricultural Districts: A Computational Approach Based on Entropy Indexes. *Environ. Impact Assess. Rev.* **2017,** *64*, 37–46.
5. Cantuarias-Villessuzanne, C.; Weigel, R.; Blain, J. Clustering of European Smart Cities to Understand the Cities' Sustainability Strategies. *Sustainability* **2021,** *13*, 513.
6. Yigitcanlar, T. Smart City Policies Revisited: Considerations for a Truly Smart and Sustainable Urbanism Practice. *World Technopolis Rev.* **2018,** *7*, 97–112.
7. Laconte, P. Smart and Sustainable Cities: What Is Smart?—What Is Sustainable? In: *International Conference on Smart and Sustainable Planning for Cities and Regions*; Springer: Cham, **2017**; pp 3–19.
8. Lee, S. Y.; Lee, J. E. Investigation on Smart City Objectives and Implications: Adaption to Silver Population in Korea as Target Citizens. *J. Korea Contents Assoc.* **2017,** *17*, 470–478.
9. Manville, C.; Cochrane, G.; Jonathan, C. A. V. E.; Millard, J.; Pederson, J. K.; Thaarup, R. K.; WiK, M. W. *Mapping Smart Cities in the EU. Mapping Smart Cities in the EU | Policy Commons*; European Parliament: Strasbourg, France, **2014**.
10. Gil-Garcia, J.R.; Helbig, N.; Ojo, A. Being Smart: Emerging Technologies and Innovation in the Public Sector. *Gov. Inf. Q.* **2014,** *31*, I1–I8.
11. Gasco-Hernandez, D. M. Is It More Than Using Data and Technology in Local Governments? Identifying Opportunities and Challenges for Cities to Become Smarter. *UMKC Law Rev.* **2017,** *85*, 915.
12. Pan, S.; Zhou, W.; Piramuthu, S.; Giannikas, V.; Chen, C. Smart City for Sustainable Urban Freight Logistics. *Int. J. Prod. Res.* **2021,** *59*, 2079–2089.
13. Bibri, S. E.; Krogstie, J. Smart Sustainable Cities of the Future: An Extensive Interdisciplinary Literature Review. *Sustain. Cities Soc.* **2017,** *31*, 183–212.
14. Kim, Y.; Choi, T.Y.; Yan, T.; Dooley, K. Structural Investigation of Supply Networks: A Social Network Analysis Approach. *J. Oper. Manag.* **2010,** *29*, 194–211.
15. Angelidou, M.; Psaltoglou, A.; Komninos, N.; Kakderi, C.; Tsarchopoulos, P.; Panori, A. Enhancing Sustainable Urban Development Through Smart City Applications. *J. Sci. Technol. Policy Manag.* **2017,** *9*, 146–169.
16. Feki, M.; Wamba, S. F.: Big Data Analytics-Enabled Supply Chain Transformation: A Literature Review. In: *49th Hawaii International Conference on System Sciences*; **2016**; pp 1123–1132.
17. Mohanty, S. P.; Choppali, U.; Kougianos, E. Everything You Wanted to Know About Smart Cities: The Internet of Things Is the Backbone. *IEEE Consum. Electron. Mag.* **2016,** *5*, 60–70.
18. Singh, A.; Jain, D.; Mehta, I.; Mitra, J.; Agrawal, S. Application of Big Data in Supply Chain Management. In: *Proceedings of the 5th International Conference of Materials Processing and Characterization (ICMPC 2016)*, Hyderabad, India, 12–13 March 2016;
19. Moktadir, M. A.; Ali, S. M.; Paul, S. K.; Shukla, N. Barriers to Big Data Analytics in Manufacturing Supply Chains: A Case Study from Bangladesh. *Comput. Ind. Eng.* **2019,** *128*, 1063–1075.

20. Nguyen, T.; Zhou, L.; Spiegler, V.; Ieromonachou, P.; Lin, Y. Big Data Analytics in Supply Chain Management: A State-of-the-Art Literature Review. *Comput. Oper. Res.* **2018,** *98*, 254–264.
21. Gandomi, A.; Haider, M. Beyond the Hype: Big Data Concepts, Methods, and Analytics. *Int. J. Inf. Manag.* **2015,** *35*, 137–144.
22. Mani, V.; Delgado, C.; Hazen, B.T.; Patel, P. Mitigating Supply Chain Risk via Sustainability Using Big Data Analytics: Evidence from the Manufacturing Supply Chain. *Sustainability* **2017,** *9*, 608.
23. Edelenbos, J.; Hirzalla, F.; van Zoonen, L.; Bouma, G.; Slob, A.; Woestenburg, A. Governing the Complexity of Smart Data Cities: Setting a Research Agenda. In: *Smart Technologies for Smart Governments*; Springer: Berlin/Heidelberg, Germany, **2017**; pp 35–54.
24. Tiwari, S.; Wee, H. M.; Daryanto, Y. Big Data Analytics in Supply Chain Management Between 2010 and 2016: Insights to Industries. *Comput. Ind. Eng.* **2018,** *115*, 319–330.
25. Ahmed, E.; Yaqoob, I.; Hashem, I. A. T.; Khan, I.; Ahmed, A. I. A.; Imran, M.; Vasilakos, A. V. The Role of Big Data Analytics in Internet of Things. *Comput. Netw.* **2017,** *129*, 459–471.
26. Ghonge, M. M.; Mane, S.; Pradhan, D. Demystifying the Role of Blockchain Technology in Healthcare and Transaction. In: *Blockchain Technologies and Applications for Digital Governance*; IGI Global, **2022**; pp 60–84.
27. Thaung, S. M.; Tun, H. M.; Win, K. K. K.; Than, M. M. Exploratory Data Analysis Based on Remote Health Care Monitoring System by Using IoT. *Communications* **2020,** *8* (1), 1–8. DOI: 10.11648/j.com.20200801.11
28. Pradhan, D.; Tun, H. M.; Dash, A. K. IoT: Security & Challenges of 5G Network in Smart Cities. *Asian J. Convergence Technol. (AJCT)* **2022,** *8* (2), 45-50.
29. Wah, N. K. S. Integration of AI/Ml in 5G Technology toward Intelligent Connectivity, Security, and Challenges. *Mach. Learn. Algorithms App. Eng.* **2023,** *239*.

CHAPTER 14

5G-Enabled Edge Computing for Smart Cities

DEVASIS PRADHAN[1], PRASANNA KUMAR SAHU[2], HLA MYO TUN[3], TIRTHA MAJUMDER[4], and RAJESWARI[1]

[1]*Department of Electronics & Communication Engineering, Acharya Institute of Technology, Bengaluru, India*

[2]*Department of Electrical Engineering, National Institute of Technology, Rourkela, Odisha, India*

[3]*Research Department, Yangon Technological University, Yangon, Myanmar*

[4]*Department of Electrical and Electronics Engineering, School of Engineering and Technology, Adamas University, Kolkata, India*

ABSTRACT

In the context of smart cities, this article examines the potential of 5G-enabled edge computing. This article discusses the expanding smart city concept and the challenges it faces in data processing, storage, and communication. The article then discusses the advantages of 5G networks and the use of edge to solve these problems. Specifically, this article examines how edge recording with 5G can support data continuity, reduce static, increase transmission speed, and improve the stability and security of smart city applications. The report also discusses 5G-enabled edge computing for traffic control, public safety, energy management, and smart city medical applications. Next, this article discusses the challenges and future research directions in 5G-enabled edge computing for smart cities.

5G Green Communication Networks for Smart Cities. Devasis Pradhan, Rajeswari, Hla Myo Tun, Naw Khu Say Wah, & Thandar Oo (Eds.)

14.1 INTRODUCTION

Edge computing powered by 5G has the potential to have a significant impact on the growth of smart cities. Shrewd urban areas depend on the assortment, handling, and investigation of a lot of information to upgrade city tasks, work on personal satisfaction for inhabitants, and improve monetary turn of events. Instead of sending data to a centralized processing location, edge computing processes data closer to where it is generated. This diminishes idleness and further develops reaction time, which is basic for applications that demand ongoing information examination and direction.[1,2]

Edge computing can be implemented on a larger scale and more effectively with 5G networks. In comparison to previous generations of wireless networks, 5G networks offer greater bandwidth, lower latency, and higher data transfer rates. This permits edge-figuring gadgets to process and communicate information all the more rapidly and proficiently. Edge computing with 5G support can be used for a wide range of tasks in smart cities. It can be used, for instance, to process data from sensors that track energy consumption, air quality, and traffic flow. In order to optimize traffic flow, reduce pollution, and conserve energy, this data can be analyzed in real time.[3]

Additionally, applications related to public safety, such as video surveillance and facial recognition, can benefit from 5G-enabled edge computing. By handling this information at the edge, policing can rapidly recognize likely dangers and answer progressively. In general, 5G-enabled edge computing is a crucial technology for smart city development. It has the potential to boost residents' quality of life, cut costs, and increase efficiency. However, it is essential to ensure that these technologies are utilized in a manner that safeguards civil liberties and privacy and that they are accessible to all community members.[4,5]

14.2 WIRELESS COMMUNICATION: EVOLUTION

From the invention of radio communication in the late 19th century to the development of modern wireless technologies like 5G, wireless communication has undergone significant change over time.[6] The following are some of the major turning points in the evolution of wireless communication:

- Radio transmission: Guglielmo Marconi came up with the idea for radio communication at the end of the 19th century, which used electromagnetic waves to send messages over long distances.
- Mobile analog telephone: The main simple versatile communication frameworks were presented during the 1970s. These frameworks permitted voice calls to be made remotely over brief distances.
- Cellular Data Networks: Digital cellular networks were introduced in the 1980s, making it possible to make voice calls wirelessly over greater distances. The first generation of cellular networks, or 1G, was built on these networks.
- 2G Systems: The subsequent age (2G) of cell networks was presented during the 1990s, which considered advanced information transmission notwithstanding voice calls. These organizations additionally presented new highlights like SMS and guest ID.
- Networks for 3G: At the beginning of the 2000s, the cellular networks of the third generation, or 3G, were introduced. These networks allowed for more advanced applications like mobile Internet access and video calling at higher data transfer rates.
- 4-G Networks: At the end of the 2000s, the cellular networks of the fourth generation (4G) were introduced. These networks allowed for more advanced applications like online gaming and high-definition video streaming at even higher data transfer rates.
- Networks for 5G: In the 2010s, the fifth generation of cellular networks, known as 5G, was introduced. Compared to previous generations, this one offers even higher data transfer rates, lower latency, and more bandwidth. This innovation empowers new applications like independent vehicles, expanded reality, and savvy urban areas.

In general, the development of wireless communication has facilitated the creation of novel technologies and applications as well as increased connectivity and communication among individuals and devices.[7,8]

14.3 EVOLUTION OF EDGE COMPUTING

In recent years, a technology known as edge computing has emerged to meet the growing demand for real-time data processing and analysis.[9–12] Here is a concise outline of the development of edge figuring:

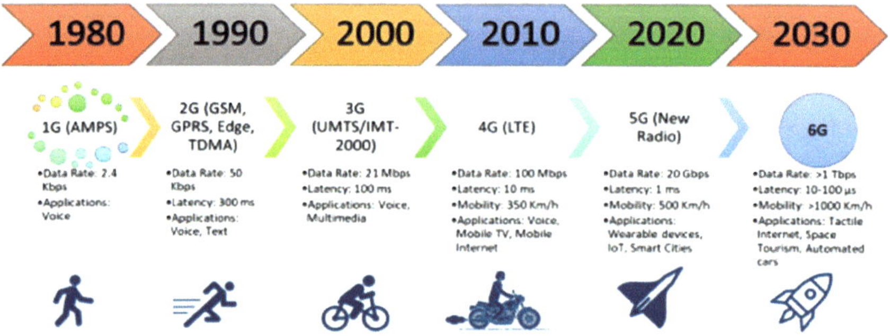

FIGURE 14.1 Evolution of wireless communication.

Source: Reprinted from Ref. [31]

- Client/Server Figuring: Applications and data were processed on centralized servers and accessed by client devices in the early days of computing. Despite its widespread use today, this architecture has bandwidth and latency limitations.
- Computing that is dispersed: Applications and data processing are moved to distributed computing environments, where data is accessed over the Internet and processing is performed on remote servers, with the growth of the Internet and cloud computing. New security and latency issues were also introduced by this model, despite its improved performance.
- Edge Technology: In response to these difficulties, edge computing was developed with the intention of processing data closer to the point where it is generated, at the "edge" of the network. This method makes it possible to analyze data in real time, increases bandwidth, and reduces latency.[10]
- Fog Computing: Fog Computing is a connected innovation that stretches out edge processing to envelop a bigger organization of gadgets, including switches, switches, and other systems administration hardware. As a result, more distributed data processing and analysis are possible.[11]
- AI at the Limits: Bringing AI processing to the edge has become a growing focus as artificial intelligence (AI) has grown in popularity. This takes into account more savvy decision-production at the edge, decreasing the need to send information to focal servers for examination.[12]

In general, the need for faster and more effective data processing and analysis has been the driving force behind the development of edge computing. Edge computing is likely to grow in importance for businesses and consumers alike as the volume of data generated by devices continues to rise.

14.4 STANDARDIZATION OF EDGE COMPUTING

In order to guarantee compatibility and interoperability among various systems and devices, standardization of edge computing is essential. Best practices and guidelines for the management and deployment of edge computing solutions can also be established with its assistance.[13,14]

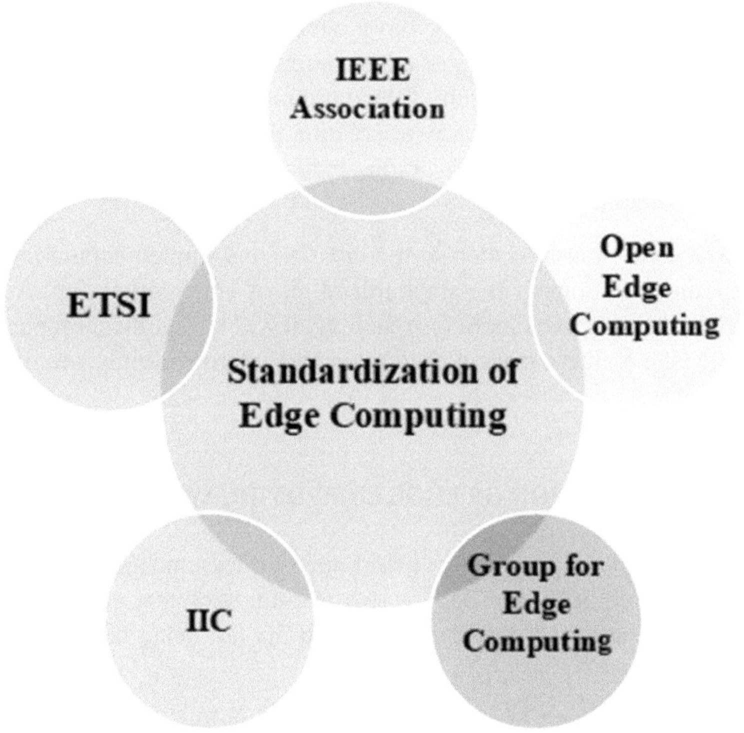

FIGURE 14.2 Edge computing standardization.

Some of the current efforts to standardize edge computing are as follows:

- IEEE Association for Standards: The IEEE Guidelines Affiliation is attempting to foster principles for edge processing, including definitions, models, and security and protection contemplations.
- Open Edge Computing (OEC): The Open Edge Computing Consortium (OEC) is a group of businesses and organizations working together to develop open frameworks and standards for edge computing. The interoperability, ease of deployment, and scalability of edge computing solutions are their objectives.[16]
- Group for Edge Computing: Companies collaborate to create edge computing standards and encourage the use of edge computing technology through the Edge Computing Consortium.[17]
- Industry IoT **Consortium (IIC)**: The IIC is a non-benefit association zeroed in on advancing the reception of the Industrial Internet of Things (IIoT). They have created a standard architecture for industrial applications of edge computing.[13,14]
- ETSI: The European Telecommunications Standards Institute (ETSI) is chipping away at creating principles for edge registering in the media communications industry, including network cutting and asset the board.

In general, a crucial step in the creation and implementation of edge computing solutions is the standardization of edge computing. As edge figuring turns out to be more far-reaching, it will be significant for associations to take on these principles to guarantee interoperability, security, and adaptability.

14.5 INTEGRATION OF EDGE COMPUTING WITH 5G NETWORK

In order to make it possible for brand-new services and applications, edge computing and 5G networks are two technologies that are increasingly being combined.[13,14] Edge computing and 5G are being combined in the following ways:

- Low Latency: For real-time applications like gaming, augmented reality, and autonomous vehicles, 5G networks offer low latency. Processing can be performed closer to the user by deploying edge computing at the network edge, thereby further reducing latency.

- Slicing the Network: Network slicing, which enables the creation of virtual networks tailored to specific applications or users, is made possible by 5G networks. These slices can benefit from individualized services and enhanced performance thanks to the availability of processing and storage resources via edge computing.[14]
- Mobile Edge Computing (MEC): MEC is a framework for 5G networks that makes edge computing possible. MEC makes it possible for applications and data to be processed with low latency by placing compute and storage resources near the edge of the network. This can work on the exhibition of uses like video web-based and gaming, and empower new applications like independent vehicles.[15]
- MEC, or multi-access edge computing: Edge computing is made possible by the MEC standard across a variety of access networks, including fixed-line, 5G, and Wi-Fi networks. MEC makes it possible for applications to move seamlessly between networks and boosts their performance by deploying edge computing resources across these networks.[16]
- Edge Cloud: Edge cloud is a term used to portray the sending of distributed computing assets at the organization's edge. Applications and data can be processed and stored closer to the user in 5G networks by deploying edge cloud resources, thereby reducing latency and improving performance.

Generally speaking, the joining of edge figuring with 5G organizations is empowering new applications and administrations that were unrealistic previously. We can anticipate the emergence of even more novel use cases and applications as these technologies continue to advance.[15–17]

14.6 SMART CITIES ENABLED WITH 5G-EDGE COMPUTING

Smart cities are urban areas that use advanced technologies to improve the quality of life of their residents, optimize resource consumption, and enhance the overall efficiency of city operations. 5G-edge computing is a technology that can enable smart cities in a number of ways.[17,18] Here are some examples:

- Intelligent Transportation Systems (ITS): 5G-edge computing can enable real-time data processing and analysis for ITS applications, such as traffic flow optimization, autonomous vehicles, and public transportation management. By deploying edge computing resources at the network edge, traffic data can be processed and analyzed in real time, allowing for better decision-making and improved traffic flow.[18]

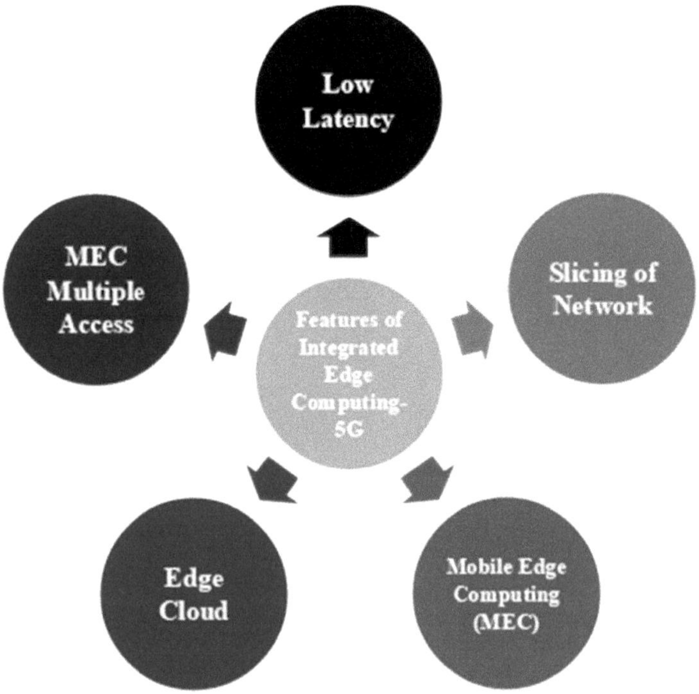

FIGURE 14.3 Integrated edge computing features.

- Public Safety and Security: 5G-edge computing can enable real-time surveillance, facial recognition, and other security applications in smart cities. By deploying edge computing resources near cameras and sensors, data can be processed and analyzed in real time, allowing for faster response times and improved security.[19]
- Energy Management: 5G-edge computing can enable energy management applications in smart cities, such as smart grid optimization, renewable energy integration, and demand response. By deploying edge computing resources near energy sensors and meters, data can

be processed and analyzed in real time, allowing for more efficient use of energy resources.[19,20]

- Environmental Monitoring: 5G-edge computing can enable environmental monitoring applications in smart cities, such as air quality monitoring, water quality monitoring, and waste management. By deploying edge computing resources near sensors and monitoring devices, data can be processed and analyzed in real time, allowing for better decision-making and improved environmental sustainability.[20]
- Healthcare: 5G-edge computing can enable remote healthcare applications in smart cities, such as telemedicine and remote patient monitoring. By deploying edge computing resources near healthcare sensors and devices, data can be processed and analyzed in real time, allowing for better healthcare outcomes and improved patient care.[21]

Overall, 5G-edge computing can enable a wide range of smart city applications, allowing for better decision-making, improved efficiency, and enhanced quality of life for city residents. As these technologies continue to evolve, we can expect to see even more innovative use cases and applications emerge.

14.7 REQUIREMENTS OF EDGE COMPUTING-ENABLED SMART CITIES

Edge computing-enabled smart cities have several requirements that need to be met to ensure that the system operates efficiently and provides the desired benefits to the residents.[21–24] Here are some of the key requirements:

- Scalability: Edge computing-enabled smart cities should be scalable to meet the growing needs of the population. This means that the system should be designed to handle a large amount of data and traffic that will be generated by different IoT devices and sensors.
- Reliability: The system should be reliable and able to operate continuously without any downtime. The reliability of the system will depend on the quality of the hardware, software, and network infrastructure deployed.
- Low Latency: The system should have low latency, which means that the data processing and analysis should be done as close to the data source as possible. This is important for applications like autonomous vehicles and real-time traffic management.

- Security: The system should be designed with security in mind to protect the data from unauthorized access and cyber-attacks. The system should include security measures like encryption, firewalls, and access control mechanisms.
- Interoperability: Edge computing-enabled smart cities should be designed with interoperability in mind, which means that the different components and systems should be able to work together seamlessly. This will require the use of standardized interfaces and protocols.
- Energy Efficiency: The system should be energy-efficient to reduce overall energy consumption and minimize the environmental impact. This can be achieved by using low-power devices, optimizing data processing, and using renewable energy sources.
- User-Centricity: The system should be designed with the needs of the residents in mind. This means that the system should be user-friendly, intuitive, and able to adapt to the changing needs of the residents.

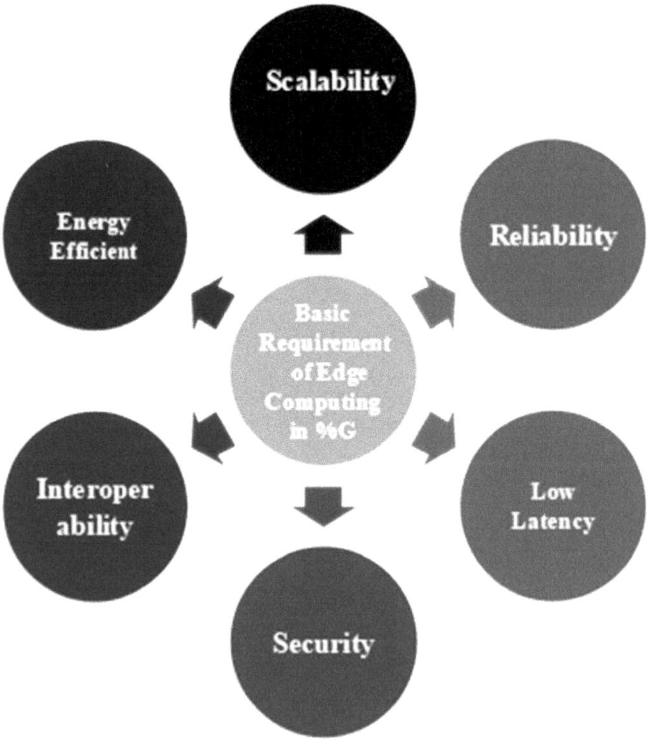

FIGURE 14.4 Requirements for edge computing.

Basically, edge computing-enabled smart cities should be designed with scalability, reliability, low latency, security, interoperability, energy efficiency, and user-centricity in mind to ensure that the system operates efficiently and provides the desired benefits to the residents.

14.8 CHALLENGES IN DEPLOYING EDGE COMPUTING SERVICES IN 5G-ENABLED CITIES

Deploying edge computing services in 5G-enabled cities can be challenging due to a number of factors. Here are some of the key challenges:

- Infrastructure: A significant amount of infrastructure, including hardware, software, and networking components, is required for the deployment of edge computing services in a city. The installation and upkeep of this can be costly and time-consuming.[25]
- Interoperability: Edge computing and 5G standards may be used by different vendors, preventing interoperability and causing compatibility issues. The integration of various systems and components within a smart city ecosystem may be difficult as a result of this.
- Privacy and security: The processing and storage of data at the network edge in edge computing can raise privacy and security concerns. It is essential to ensure that data is safe and protected from unauthorized access, particularly in sensitive applications like healthcare and public safety.[24,25]
- Network Limit: 5G organizations offer high transfer speed and low dormancy; however, they likewise require huge organization ability to deal with a lot of information produced by edge-registering applications. It is essential to check that the network infrastructure can handle the increased demand.[26]
- Workforce and skillset: Data analysis, cybersecurity, and network architecture are just a few of the specialized skills and knowledge required for edge computing. It can be difficult to ensure that these systems are installed and maintained by a skilled workforce.
- Policies and Regulations: Edge computing services can also be impacted by policies and regulations in cities with 5G. Data ownership, data privacy, and liability are examples of these.

In general, conveying edge registering administrations in 5G-empowered urban areas requires cautious preparation, foundation advancement, and consideration regarding security and protection concerns. To realize these technologies' full potential in smart city applications, it will be necessary to address these obstacles.

14.9 MITIGATE THE CHALLENGES IN DEPLOYING EDGE COMPUTING

There are several ways to mitigate the challenges of deploying edge computing in smart cities. Overcoming the challenges of deploying edge computing in smart cities will require a coordinated effort between different stakeholders, including government agencies, industry partners, and academic institutions.[26,27] By implementing effective planning, standardization, security, network capacity, skillset and workforce development, regulations and policies, and pilot projects, it is possible to overcome these challenges and realize the full potential of edge computing in smart cities. Here are some of the strategies:

- Planning and Collaboration: Effective planning and collaboration between different stakeholders, including government agencies, industry partners, and academic institutions, can help ensure that the deployment of edge computing in smart cities is well coordinated and addresses the needs of all stakeholders.
- Standardization: Standardization of edge computing and 5G technologies can help ensure interoperability and compatibility between different systems and components. This can help reduce costs and improve efficiency.
- Security and Privacy: Security and privacy should be a top priority when deploying edge computing in smart cities. Implementing strong security measures, such as encryption, access control, and monitoring, can help protect data and mitigate security threats.[28]
- Network Capacity: Ensuring that the network infrastructure can handle the increased demand for data processing and analysis is critical. This can be achieved by deploying a combination of wired and wireless networks, optimizing network design and deployment, and using advanced network management tools.[29]

- Skillset and Workforce: Building a skilled workforce that can design, deploy, and maintain edge computing systems is essential. This can be achieved through training programs, partnerships with educational institutions, and collaboration with industry partners.
- Regulations and Policies: Governments can play a critical role in mitigating challenges by establishing policies and regulations that support the deployment of edge computing in smart cities. This can include measures to ensure data privacy, data ownership, and liability.[30]
- Pilot Projects: Conducting pilot projects can help identify and address potential challenges before deploying the technology on a larger scale. This can also help build public trust and support for the technology.

14.10 CONCLUSION AND FUTURE PERSPECTIVE

In conclusion, 5G-Edge computing-enabled smart cities have the potential to transform urban living and provide residents with a more efficient, sustainable, and connected environment. By combining the benefits of 5G and edge computing technologies, smart cities can achieve low latency, high bandwidth, and efficient data processing and analysis, which are essential for applications like autonomous vehicles, smart traffic management, and public safety. However, deploying edge computing in smart cities also poses several challenges, including network capacity, security, interoperability, and workforce development. To overcome these challenges, it is essential to implement effective planning, collaboration, standardization, security, network optimization, and skillset development strategies. The future of 5G-Edge computing-enabled smart cities looks promising, with continued innovation in technology and the increasing adoption of IoT devices, sensors, and AI. As the world's population continues to grow, urban areas will face increasing pressure to address issues such as congestion, pollution, and public safety. Smart cities enabled by 5G-Edge computing will play a critical role in addressing these challenges and providing residents with a better quality of life. 5G-Edge computing-enabled smart cities represent a significant opportunity to create sustainable and livable urban environments for the future. By addressing the challenges and leveraging the benefits of this technology, we can create a more connected, efficient, and sustainable future for all.

KEYWORDS

- **real-time data processing**
- **5G**
- **edge computing**
- **smart cities**
- **latency**
- **bandwidth**
- **reliability**
- **security**
- **traffic management**
- **energy management**
- **healthcare**

REFERENCE

1. Ahmed, E.; Yaqoob, I.; Gani, A.; Imran, M.; Guizani, M. Internet-of Things-Based Smart Environments: State of the Art, Taxonomy, and Open Research Challenges. *IEEE Wireless Commun.* **2016,** *23* (5), 10–16.
2. Xie, J.; Tang, H.; Huang, T.; Yu, F. R.; Xie, R.; Liu, J.; Liu, Y. A Survey of Blockchain Technology Applied to Smart Cities: Research Issues and Challenges, in the Press. *IEEE Commun. Surveys Tutorials* **2019**.
3. Eckhoff, D.; Wagner, I. Privacy in the Smart City—Applications, Technologies, Challenges, and Solutions. *IEEE Commun. Surveys Tutorials* **2018,** *20* (1), 489–516.
4. Ianuale, N.; Schiavon, D.; Capobianco, E. Smart Cities, Big Data, and Communities: Reasoning from the Viewpoint of Attractors. *IEEE Access* **2016,** *4*, 41–47.
5. Rostirolla, G.; Righi, R. d. R.; Barbosa, J. L. V.; da Costa, C. A. Elcity: An Elastic Multilevel Energy Saving Model for Smart Cities. *IEEE Trans. Sustain. Comput.* **2018,** *3* (1), 30–43.
6. Ramaprasad, A.; Sanchez-Ortiz, A.; Syn, T. A Unified Definition of a Smart City. In: *Springer International Conference on Electronic Government*; St. Petersburg, Russia, **2017**; pp 13–24.
7. Mehmood, Y.; Ahmad, F.; Yaqoob, I.; Adnane, A.; Imran, M.; Guizani, S.; "Internet-of-Things-Based Smart Cities: Recent Advances and Challenges. *IEEE Commun. Mag.* **2017,** *55* (9), 16–24.
8. Buyya, R.; Yeo, C. S.; Venugopal, S.; Broberg, J.; Brandic, I. Cloud Computing and Emerging It Platforms: Vision, Hype, and Reality for Delivering Computing as the 5th Utility. *Fut. Gen. Comput. Syst.* **2009,** *25* (6), 599–616.

9. Bilal, K.; Malik, S. U. R.; Khan, S. U.; Zomaya, A. Y. Trends and Challenges in cloud Datacenters. *IEEE Cloud Computing* **2014,** *1* (1), 10–20.
10. Lehrig, S.; Eikerling, H.; Becker, S. Scalability, Elasticity, and Efficiency in Cloud Computing: A Systematic Literature Review of Definitions and Metrics. In *Proceedings of the 11th ACM International Conference on Quality of Software Architectures*, Montreal, QC, Canada, **2015**; pp 83–92.
11. Liu, Y.; Peng, M.; Shou, G.; Chen, Y.; Chen, S. Toward Edge Intelligence: Multi-Access Edge Computing for 5G and Internet of Things. *IEEE Internet of Things J.* **2020,** *7* (8), 6722–6747.
12. Javed, A. R.; Ahmed, W.; Pandya, S.; Maddikunta, P. K.; Alazab, M.; Gadekallu, T. R. A Survey of Explainable Artificial Intelligence for smart cities. *Electronics*. **2023,** *12* (4), 1020.
13. Baldi, G.; Megaro, A.; Carrubbo, L. Small-Town Citizens' Technology Acceptance of Smart and Sustainable City Development. *Sustainability* **2022,** *15* (1), 325.
14. Ali, S.; Armand, T. P.; Athar, A.; Hussain, A.; Ali, M.; Yaseen, M.; Joo, M. I.; Kim, H. C. Metaverse in Healthcare Integrated with Explainable AI and Blockchain: Enabling Immersiveness, Ensuring Trust, and Providing Patient Data Security. *Sensors*. **2023,** *23* (2), 565.
15. Segura-Garcia, J.; Perez-Solano, J. J.; Felici-Castell, S.; Montoya-Belmonte, J.; Lopez-Ballester, J.; Navarro, J. M. Sustainable Soundscape Monitoring of Modified Psycho-Acoustic Annoyance Model with Edge Computing for 5G IoT Systems.
16. Chhabra, S.; Aiden, M. K.; Sabharwal, S. M.; Al-Asadi, M. 5G and 6G Technologies for Smart City. In: *Enabling Technologies for Effective Planning and Management in Sustainable Smart Cities*; Springer International Publishing: Cham, **2023**; pp 335–365.
17. Rajawat, A. S.; Goyal, S. B.; Bedi, P.; Verma, C.; Ionete, E. I.; Raboaca, M. S. 5G-Enabled Cyber-Physical Systems for Smart Transportation Using Blockchain Technology. *Mathematics* **2023,** *11* (3), 679.
18. Singh, S. 5G Enabled Network Technology Trends for Smart Healthcare Systems. *5G Wireless Commun. Syst. Healthcare Inf.* **2023**, 29–43.
19. Hassan, M. A.; Javed, R.; Granelli, F.; Gen, X.; Rizwan, M.; Ali, S. H.; Junaid, H.; Ullah, S. Intelligent Transportation Systems in Smart City: A Systematic Survey. In: *2023 International Conference on Robotics and Automation in Industry (ICRAI)*, **2023** Mar 3; IEEE, 2023; pp 1–9.
20. Jeyakumar, V.; Abirami, K. R.; Saraswathi, S.; Kumaran, R. S.; Marthi, G. Secure Medical Image Storage and Retrieval for Internet of Medical Imaging Things Using Blockchain-Enabled Edge Computing. In: *Intelligence Edge Computing for Cyber-Physical Applications*; Academic Press, **2023**; pp 85–110.
21. Wah, N. K. Integration of AI/Ml in 5G Technology Toward Intelligent Connectivity, Security, and Challenges. *Mach. Learn. Algorithms App. Eng*. **2023,** *239.*
22. Priyanka, K.; Mallavaram, G.; Raj, A.; Pradhan, D. Cognitiveness of 5G Technology Toward Sustainable Development of Smart Cities. *Decis. Support Syst. Smart City App*. **2022,** 189–203.
23. Pradhan, D.; Ray, S.; Dash A. A Critical Review on Sustainable Development of Green Smart Cities (GSCs) for Urbanization. *J. Altern. Renew. Energy Sources* **2022,** *8*, 21–29. http://dx.doi.org/10.46610/JOARES.2022.v08i03.003

24. Pradhan, D.; Dash, A.; Tun, H. M.; Wah, N. K. S.; Oo T. A Sustainable Key Enabler for mm-Wave Beamforming in 5G Environment. *J. VLSI Design Signal Process.* **2022,** *8*, pp 10–17. http://dx.doi.org/10.46610/JOVDSP.2022.v08i03.002
25. Dash, A.; Pradhan, D.; Tun, H. M.; Naing, Z. M. Integration of AI to Enhance 5G Capabilities in Smart Cities. *J. Image Process. Artif. Intell.* **2022,** *8*, 14–20. http://dx.doi.org/10.46610/JOIPAI.2022.v08i03.003
26. Dash, A.; Pradhan, D.; Tun, H. M.; Naing, Z. M. m-MTC for Optimized Communication in 5G. *J. Netw. Secur. Comput. Netw.* **2022,** *8*, 1–8. http://dx.doi.org/10.46610/JONSCN.2022.v08i03.001
27. Pradhan, D.; Dash, A.; Tun, H. M.; Wah, N. K. S.; Oo, T. Improvement of Capacity and QoE: Distributed Massive MIMO (DM-MIMO) Technology-5G. *J. Netw. Secur. Comput. Netw.* **2022,** *8*, 9–17. http://dx.doi.org/10.46610/JONSCN.2022.v08i03.002
28. Sinha, H. K.; Kumar, A.; Pradhan, D. A Study of Various Peak to Average Power Ratio (PAPR) Reduction Techniques for 5G Communication System (5G-CS). In: *Optimization Techniques in Engineering*; Wiley. **2023**; pp. 437–454. http://dx.doi.org/10.1002/9781119906391.ch27
29. Pradhan, D.; Tun, H. M.; Dash, A. K. IoT: Security & Challenges of 5G Network in Smart Cities. *Asian J. Convergence Technol. (AJCT)* **2022,** *8* (2), 45–50.
30. Pradhan, D.; Sahu, P. K.; Goje, N. S.; Ghonge, M. M.; Tun, H. M.; Rajeswari, R.; Pramanik, S. Security, Privacy, Risk, and Safety Toward 5G Green Network (5G-GN). *Cyber Secur. Netw. Secur.* **2022,** 193–216.
31. Tufail, Ali & Namoun, Abdallah & Alrehaili, Ahmed & Ali, Arshad. A Survey on 5G Enabled Multi-Access Edge Computing for Smart Cities: Issues and Future Prospects. 2021. 21. 107. 10.22937/IJCSNS.2021.21.6.15.

CHAPTER 15

Secure M2M Communication for 5G-enabled Cities with Its Challenges

DEVASIS PRADHAN[1], HLA MYO TUN[2], ZAW MIN NAING[3], THANDAR OO[4], and NAW KHU SAY WAH[4]

[1]*Department of Electronics & Communication Engineering, Acharya Institute of Technology, Bengaluru, India*

[2]*Research Department, Yangon Technological University, Yangon, Myanmar*

[3]*Department of Research and Innovation, Ministry of Science and Technology, Yangon, Myanmar*

[4]*Department of Electronic Engineering, Yangon Technological University, Yangon, Myanmar*

ABSTRACT

The advent of 5G technology has opened up new possibilities for machine-to-machine (M2M) communication in smart cities. However, as the number of connected devices and applications increases, so do the security challenges associated with M2M communication. This paper provides an abstract overview of the challenges faced by 5G-enabled cities in securing M2M communication. The paper starts by introducing the concept of M2M communication and its importance in smart cities. It then discusses the security challenges faced by M2M communication in 5G-enabled cities, such as device authentication, confidentiality, integrity,

5G Green Communication Networks for Smart Cities. Devasis Pradhan, Rajeswari, Hla Myo Tun, Naw Khu Say Wah, & Thandar Oo (Eds.)

and availability. The paper also highlights the importance of privacy and data protection in M2M communication. Finally, the paper concludes by discussing the need for effective security measures to ensure secure M2M communication in 5G-enabled cities.

15.1 INTRODUCTION

M2M (machine-to-machine) communication refers to direct communication between devices without human intervention. 5G-enabled cities are those that leverage the capabilities of 5G networks to improve the efficiency, safety, and quality of life of their citizens. In such cities, M2M communication can play a crucial role in enabling various applications and services. For instance, M2M communication can facilitate the deployment of smart transportation systems, which can help reduce congestion, optimize routes, and enhance safety.[1] Additionally, M2M communication can enable the deployment of smart grids, which can help improve energy efficiency and reduce costs. 5G networks provide several advantages for M2M communication. First, 5G networks offer high bandwidth and low latency, which can support the real-time transmission of data between devices. Second, 5G networks offer high reliability, which is essential for applications that require uninterrupted connectivity.[2,3] Finally, 5G networks support massive IoT (Internet of Things) deployments, which can enable the connection of millions of devices to the network. Overall, M2M communication can play a critical role in the development of 5G-enabled cities. By enabling the deployment of smart systems and services, M2M communication can help improve the quality of life for citizens and enhance the efficiency of city operations.

15.2 OVERVIEW OF 5G

5G is the fifth generation of cellular networks, succeeding 4G/LTE. It is designed to provide faster data rates, lower latency, and higher capacity than its predecessors. 5G networks are expected to revolutionize the way we connect and communicate by enabling new applications and services that were not possible with previous generations of wireless networks. Figure 15.1 shows the supportive features of 5G network.

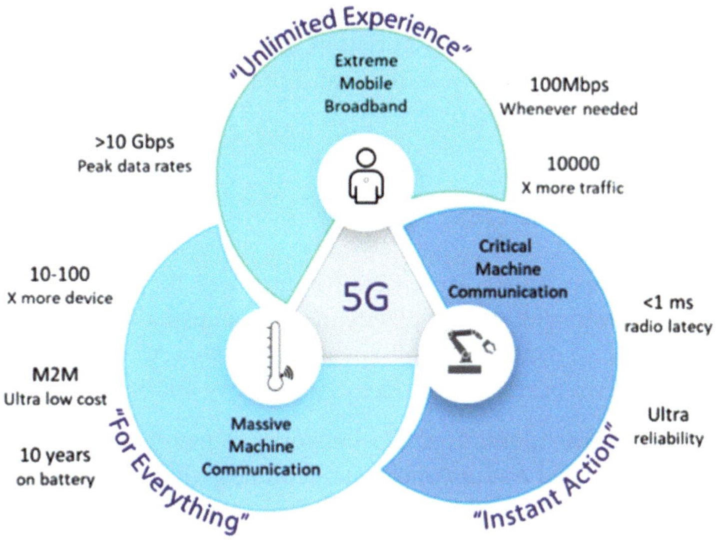

FIGURE 15.1 Basic features supported by 5G network

Source: Reprinted from 5GZONE. https://the5gzone.com/index.php/introduction-to-5g/

Some of the key features of 5G include:

- High data rates: 5G networks can provide data rates up to 20 Gbps, which is about 20 times faster than 4G/LTE.
- Low latency: 5G networks can have latency as low as 1 ms, which is critical for applications that require real-time communications, such as virtual and augmented reality, autonomous vehicles, and remote surgery.
- Massive connectivity: 5G networks can support up to 1 million devices per square kilometer, which is essential for the Internet of Things (IoT) and the deployment of smart cities.
- Network slicing: 5G networks can be divided into virtual networks that are customized for different applications, allowing for more efficient use of network resources.
- Edge computing: 5G networks can enable computing and storage resources to be moved closer to the end-user, reducing latency and improving application performance.
- Security: 5G networks have stronger security features than previous generations, including improved authentication and encryption protocols.

5G networks are expected to have a significant impact on various industries, such as healthcare, transportation, and manufacturing, by enabling new applications and services that can improve efficiency, productivity, and quality of life.

15.3 5G SMART CITIES

5G-enabled smart cities are cities that leverage the capabilities of 5G networks to improve the efficiency, safety, and quality of life of their citizens.[8–10] By enabling high-speed, low-latency connectivity, 5G networks can support a wide range of smart city applications, including:

- **Smart transportation:** 5G networks can support the deployment of connected and autonomous vehicles, which can reduce traffic congestion, improve safety, and increase efficiency. Additionally, 5G networks can enable real-time traffic monitoring and management, as well as smart parking systems.
- **Smart energy:** 5G networks can support the deployment of smart grid systems, which can improve energy efficiency, reduce costs, and enable the integration of renewable energy sources.
- **Smart healthcare:** 5G networks can support the deployment of remote healthcare services, such as telemedicine and remote monitoring, which can improve access to healthcare and reduce costs.
- **Smart buildings:** 5G networks can enable the deployment of smart building systems, which can improve energy efficiency, optimize space utilization, and enhance occupant comfort and safety.
- **Smart public safety:** 5G networks can enable real-time monitoring and response to emergencies, as well as the deployment of smart surveillance systems and predictive policing.

Overall, 5G-enabled smart cities can offer a range of benefits, including improved efficiency, increased safety, and enhanced quality of life for their citizens. However, the deployment of 5G networks and smart city infrastructure also raises concerns around privacy, security, and equity, which must be addressed to ensure that the benefits of 5G-enabled smart cities are accessible to all citizens.[10–12]

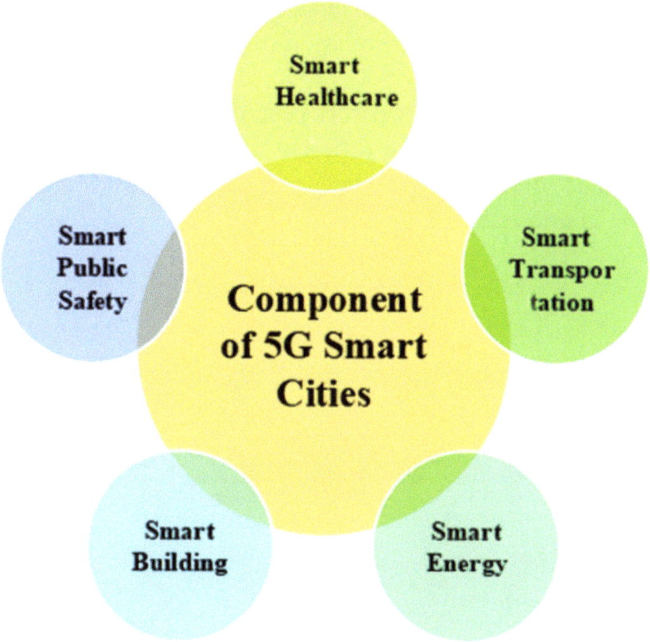

FIGURE 15.2 A basic component of smart cities enabled 5G network.

15.4 M2M COMMUNICATION IN 5G NETWORK

M2M (machine-to-machine) communication is an important aspect of 5G networks, as it enables direct communication between devices without human intervention.[12–15] M2M communication is essential for the Internet of Things (IoT) and enables a wide range of applications, including smart homes, smart cities, and industrial automation.

In 5G networks, M2M communication can leverage the following features:

- **Massive connectivity:** 5G networks can support up to 1 million devices per square kilometer, which is essential for M2M communication and the deployment of large-scale IoT applications.
- **Low latency:** 5G networks can have latency as low as 1 ms, which is critical for real-time M2M communication and applications that require an immediate response.

- **High bandwidth:** 5G networks can provide high bandwidth, which is essential for the transmission of large amounts of data between devices.
- **Network slicing:** 5G networks can be divided into virtual networks that are customized for different applications, enabling more efficient use of network resources and ensuring that M2M communication is prioritized.

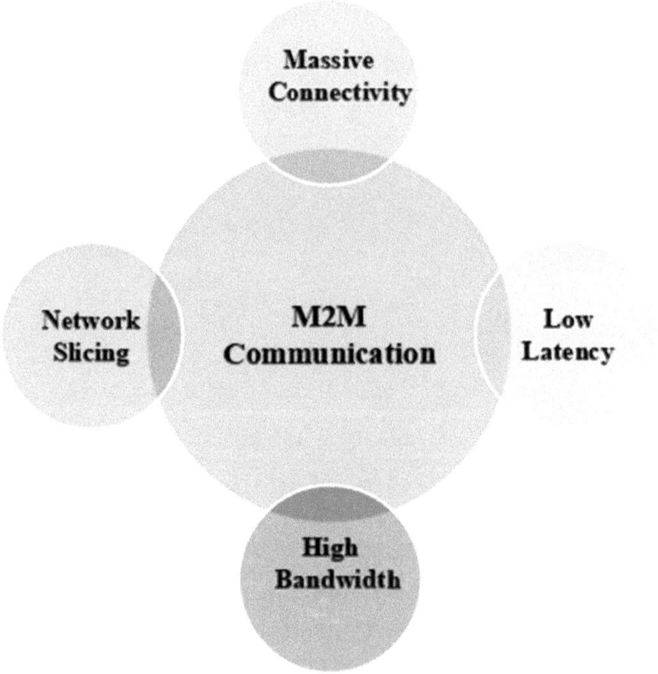

FIGURE 15.3 Basic features of M2M communication.

15.5 RELATIONSHIP BETWEEN M2M AND IOT

M2M (machine-to-machine) communication and IoT (Internet of Things) are closely related but not interchangeable terms. M2M refers to the communication between two or more machines, devices, or sensors without human intervention. It involves the exchange of data between machines or devices, typically through wired or wireless networks. M2M communication has been used for years in industrial and manufacturing settings, such as in factories or oil refineries, to enable automated processes.[15–18] On the other hand, IoT refers to a network of physical devices, vehicles, home appliances, and other items that are embedded with sensors, software, and connectivity, which enables them to connect and exchange data with other devices and systems over the Internet. IoT systems may involve M2M communication, but they also involve the interaction between machines and humans, and the ability to remotely control devices and systems. Therefore, while M2M is a subset of IoT, IoT encompasses a broader range of technologies and applications that go beyond the simple communication between machines. Table 15.1 discusses the comparison between M2m and IoT based on a certain parametric study.

TABLE 15.1 Comparison Between M2M and IoT.

Sl. No.	Parameter	IoT	M2M
1	Basic Definition	IoT (Internet of Things) refers to a network of physical objects or "things" that are embedded with sensors, software, and connectivity that enables them to connect and exchange data with other devices or systems over the Internet.	M2M technology enables devices to interact with each other, share data, and perform tasks without the need for human intervention, thereby streamlining processes and increasing efficiency.
2	Type of connectivity	Through IP network	Point-to-point
3	Protocols	Internet Protocol (IP)	Old proprietary protocols and communication techniques
4	Data sharing	Data collected is shared with other applications	Data collected is not shared with other applications
5	Major area focused	To address the everyday needs of humans	For monitoring and control of 1 or few infrastructure/ assets.
6	Dependency	Devices rely on Internet	Do not rely on the Internet
7	Scope of device connected	Large no. of devices connected	Limited devices connected
8	Scalability	More scalable due to cloud-based services	Less scalable as compared to IoT

15.6 M2M COMMUNICATION ARCHITECTURE

M2M (machine-to-machine) communication architecture typically consists of three layers: the device layer, the network layer, and the application layer.

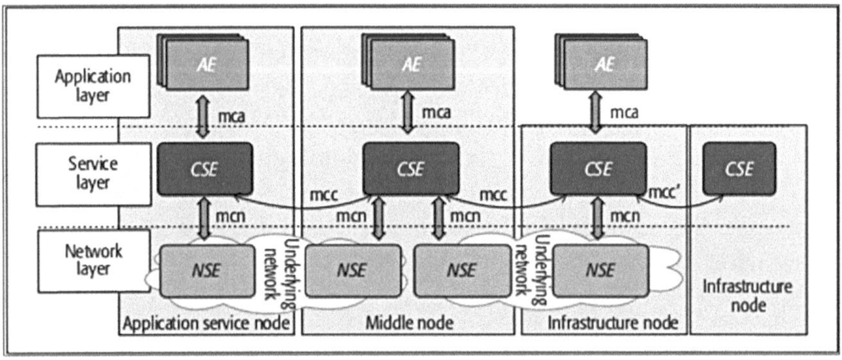

FIGURE 15.4 The architecture of M2M communication.

Source: Adapted from Ref.[8]

- **Device Layer:** The device layer includes all of the physical devices that make up the M2M system, such as sensors, actuators, and other types of devices that are used to collect data, monitor conditions, and control processes. These devices are connected to the network and communicate with each other and with other layers in the M2M architecture. In addition to the physical devices, the device layer includes the software and firmware that control the behavior of these devices. This software is responsible for collecting data from the sensors, processing that data, and sending it to other layers in the M2M architecture. The firmware is responsible for controlling the behavior of the devices themselves, such as setting thresholds and triggering actions based on sensor readings.
- **Network Layer:** The network layer includes the network infrastructure and protocols that are used to connect the devices in the M2M system. This includes wired and wireless communication technologies, such as Ethernet, Wi-Fi, Bluetooth, and cellular networks. The network layer also includes the protocols used to exchange data between devices, such as MQTT, CoAP, and HTTP. One of the key functions of the network layer is to ensure reliable and secure

communication between devices. This includes mechanisms for data encryption and authentication to protect against unauthorized access and data tampering. It also includes mechanisms for error detection and correction to ensure that data is transmitted and received correctly. Another important function of the network layer is to enable communication between devices that may be located in different geographical locations or connected to different types of networks. This is achieved through the use of gateways and other devices that can translate between different network protocols and enable communication between devices that would otherwise be incompatible.

- **Application Layer:** The application layer includes the software and services that are used to process and analyze the data collected by the devices in the M2M system. This includes data processing and storage, analytics and visualization, and application integration and APIs. One of the key functions of the application layer is to transform raw data collected from the devices into actionable insights that can be used to improve business processes and drive value for the organization. This involves processing and aggregating data, applying analytics and machine learning algorithms, and generating reports and dashboards to provide visibility into the performance of the M2M system. Another important function of the application layer is to enable integration with other systems and applications. This is achieved through the use of APIs and other integration mechanisms that allow data to be shared between the M2M system and other systems, such as ERP, CRM, and other enterprise applications.

Basically, the M2M communication architecture provides a scalable and flexible framework for connecting and managing devices, enabling businesses and organizations to automate processes, reduce costs, and improve efficiency.[18–22]

15.7 STANDARDIZATION OF M2M COMMUNICATION

Standardization plays a critical role in the development and deployment of M2M (machine-to-machine) communication solutions. Standardization ensures interoperability, facilitates integration between devices and systems, and helps to reduce costs and time-to-market.[20–23] Standardization efforts in M2M are ongoing, and new standards and technologies are being developed to address emerging use cases and challenges in M2M communication.

Some several organizations and bodies work on developing M2M standards, including:

- International Organization for Standardization (ISO): ISO has developed several standards related to M2M, including ISO/IEC 30141, which provides a framework for M2M service architecture and application program interfaces (APIs).
- European Telecommunications Standards Institute (ETSI): ETSI has developed several standards related to M2M, including the one M2M standard, which provides a common platform for M2M communications and supports interoperability between different M2M systems and applications.
- Institute of Electrical and Electronics Engineers (IEEE): IEEE has developed several standards related to M2M, including IEEE 802.15.4, which defines the physical and media access control (MAC) layer specifications for low-rate wireless personal area networks (LR-WPANs).
- International Telecommunication Union (ITU): ITU has developed several standards related to M2M, including ITU-T Y.2060, which provides a conceptual framework for M2M communication.
- Open Mobile Alliance (OMA): OMA has developed several standards related to M2M, including Lightweight M2M (LwM2M), which provides a device management protocol for M2M applications.

15.8 M2M APPLICATION AREA

M2M (Machine-to-Machine) communication refers to the direct communication between machines, devices, and sensors without the need for human intervention. Figure 15.4 discusses the area of application intended for M2M communication.

The application areas for M2M communication are broad and varied and include:

- **Industrial Automation:** M2M (Machine-to-Machine) communication plays a critical role in industrial automation. It enables machines and devices to communicate with each other and exchange data in real time, without the need for human intervention. This

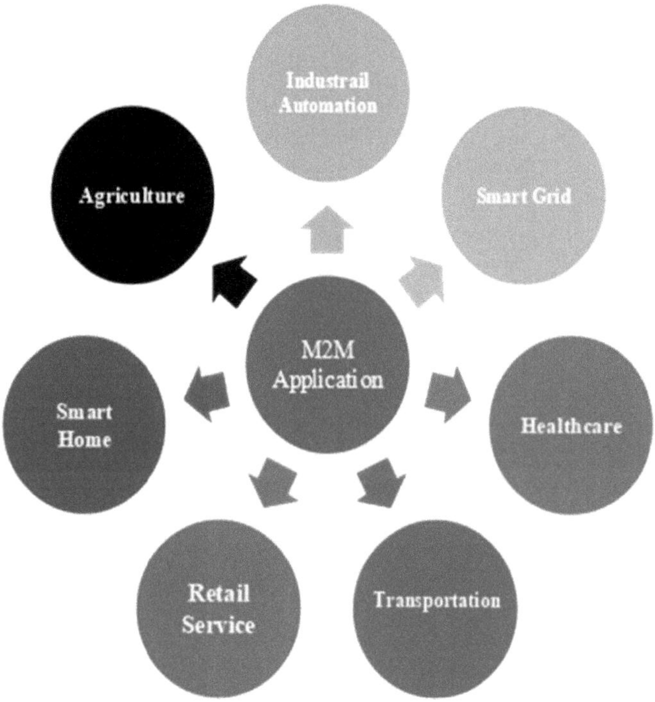

FIGURE 15.5 Area of application.

communication allows for increased efficiency, reduced downtime, and improved quality control in the manufacturing process. One example of M2M industrial automation is in a manufacturing plant, where machines are connected to a central control system. The machines can communicate with each other, and with the control system, to optimize production processes. For example, if one machine experiences a malfunction, it can automatically alert other machines in the system to adjust their operations accordingly. This reduces the risk of downtime and minimizes the impact on production schedules. M2M communication also allows for better quality control in industrial automation. Machines can communicate with each other to ensure that they are producing products that meet specific quality standards. For example, sensors on a production line can monitor the weight and dimensions of each product and adjust the manufacturing process to ensure that each product meets the required specifications. In addition, M2M communication can help to reduce waste in the manufacturing process. Machines can communicate with each other to optimize material usage and reduce the amount of waste generated. This can help to save costs and reduce the environmental impact of manufacturing operations. Overall, M2M industrial automation is a powerful tool that can help manufacturers to increase efficiency, reduce downtime, improve quality control, and reduce waste. As technology continues to evolve, we will likely see even more innovative uses for M2M communication in industrial automation.

- **Smart Grid:** M2M (Machine-to-Machine) communication plays a critical role in the development of smart grids. A smart grid is an advanced electricity grid that uses advanced technologies to improve efficiency, reliability, and sustainability. M2M communication is an essential component of a smart grid because it allows devices and sensors to communicate with each other in real time and enables the grid to respond dynamically to changes in demand and supply. One of the key benefits of M2M communication in a smart grid is the ability to monitor and manage the grid in real time. Sensors and devices can provide real-time data on electricity usage, demand, and supply, allowing grid operators to optimize the grid's performance and respond quickly to changes in demand or supply. M2M communication can also enable better integration of renewable energy sources into the grid. Renewable energy sources, such as solar and wind, are often intermittent and can be difficult to integrate into traditional electricity grids. However, M2M communication can help to manage the variability of renewable energy sources by adjusting the grid in real time to balance supply and demand. Another benefit of M2M communication in a smart grid is the ability to improve the reliability and security of the grid. M2M communication allows for real-time monitoring of the grid's infrastructure and can alert operators to potential problems before they become significant issues. This can help to prevent power outages and other disruptions to the grid. Finally, M2M communication can also help to reduce costs and improve efficiency in the operation of the grid. By providing real-time data on energy usage and supply, M2M communication can enable more accurate forecasting and planning of grid operations, reducing the need for expensive infrastructure upgrades and improving overall efficiency.

- **Healthcare:** M2M (Machine-to-Machine) communication has the potential to revolutionize healthcare by enabling remote monitoring and real-time data sharing between patients and healthcare providers. This technology can help to improve patient outcomes, reduce healthcare costs, and increase access to care. One of the primary uses of M2M communication in healthcare is remote patient monitoring. Devices and sensors can be used to monitor patients' vital signs, such as heart rate, blood pressure, and blood glucose levels, and transmit this data to healthcare providers in real time. This can allow healthcare providers to monitor patient's health remotely, identify potential problems early, and intervene before they become serious issues. M2M communication can also be used to improve medication adherence, which is a significant problem in healthcare. Devices can be used to remind patients to take their medications, monitor medication usage, and alert healthcare providers if a patient misses a dose or is experiencing adverse side effects. Another area where M2M communication can be useful is in telemedicine. Patients can use video conferencing and other technologies to communicate with healthcare providers remotely, reducing the need for in-person visits and increasing access to care, particularly in rural or underserved areas. Finally, M2M communication can also help to improve healthcare outcomes by enabling healthcare providers to share data more easily. For example, healthcare providers can use M2M communication to share patient data, such as medical records, test results, and imaging studies, securely and efficiently. This can help to ensure that patients receive the best possible care, regardless of their location or healthcare provider.
- **Transportation:** M2M (Machine-to-Machine) communication plays a critical role in the transportation industry by enabling real-time data sharing between vehicles, infrastructure, and fleet managers. This technology can help to improve safety, and efficiency, and reduce operating costs. One of the primary uses of M2M communication in transportation is in telematics. Telematics is the use of technology to monitor vehicles' location, speed, and other performance metrics in real time. This data can be used to optimize routes, reduce fuel consumption, and improve safety by identifying unsafe driving

behaviors. M2M communication can also be used to improve maintenance and reduce downtime. Sensors and devices can be used to monitor vehicles' performance, identify potential maintenance issues, and alert fleet managers before they become significant problems. This can help to reduce downtime and prevent costly repairs. Another area where M2M communication can be useful is logistics and supply chain management. M2M communication can enable real-time tracking of shipments, providing visibility into the location and status of goods at all times. This can help to improve delivery times, reduce the risk of theft or damage to goods, and optimize supply chain operations. M2M communication can also be used to improve safety in transportation. For example, sensors and devices can be used to monitor road conditions and alert drivers to potential hazards, such as black ice or construction zones. M2M communication can also be used to enable vehicle-to-vehicle communication, allowing vehicles to share information about their location, speed, and other performance metrics to help prevent accidents and improve traffic flow. Overall, M2M communication has the potential to transform the transportation industry by improving safety, and efficiency, and reducing operating costs. As technology continues to evolve, we are likely to see even more innovative uses for M2M communication in transportation.

- **Retail:** M2M (Machine-to-Machine) communication has the potential to revolutionize the retail industry by enabling real-time data sharing between retailers, customers, and inventory management systems. This technology can help to improve customer experiences, increase efficiency, and reduce costs. One of the primary uses of M2M communication in retail is in inventory management. Sensors and devices can be used to monitor inventory levels in real time, enabling retailers to optimize inventory levels, reduce waste, and avoid stockouts. This can help to ensure that customers can find the products they need, when they need them, and improve customer satisfaction. M2M communication can also be used to enable more personalized customer experiences. For example, retailers can use customer data, such as purchase history and preferences, to offer personalized recommendations and promotions in real time. This can help to improve customer loyalty and increase sales. Another area where M2M communication can be useful is in supply chain management. M2M communication can enable real-time tracking of shipments, providing visibility into the location and status of goods at all times. This can help to improve delivery times, reduce the risk of theft or damage to goods, and optimize supply chain operations. M2M communication can also be used to improve store operations. For example, retailers can use sensors and devices to monitor foot traffic in stores, identify peak shopping times, and adjust staffing levels accordingly. This can help to improve customer experiences, reduce wait times, and improve efficiency. Finally, M2M communication can also help to reduce costs in the retail industry. By enabling real-time monitoring and management of inventory and operations, retailers can reduce waste, optimize operations, and increase efficiency, reducing operating costs and improving profitability.

- **Smart Homes:** M2M (Machine-to-Machine) communication has the potential to transform the way we live in our homes by enabling real-time data sharing between devices and appliances. This technology can help to increase convenience, improve energy efficiency, and enhance security. One of the primary uses of M2M communication in smart homes is in home automation. M2M communication can enable devices and appliances to communicate with each other and with the homeowner, allowing for remote control of lights, thermostats, security systems, and other home features. This can help to increase convenience and improve energy efficiency by allowing homeowners to control their homes from anywhere, at any time. M2M communication can also be used to improve energy efficiency in smart homes. Sensors and devices can be used to monitor energy usage in real time, enabling homeowners to identify areas where they can reduce their energy consumption and save money on their utility bills. Smart home devices can also be programmed to adjust energy usage based on occupancy, time of day, and other factors, further improving energy efficiency. Another area where M2M communication can be useful is in home security. M2M communication can enable real-time monitoring of security systems, allowing homeowners to receive alerts and notifications if there is a security breach or other issues. Smart home devices, such as door locks and cameras, can also be controlled remotely, providing an extra layer of security and peace of mind. M2M communication can also be used to improve home health and wellness. For example, sensors and devices can be used to monitor air quality, humidity levels, and other environmental factors, providing homeowners with the information they need to improve their health and well-being. Smart home devices can also be used to control and automate home healthcare devices, such as blood pressure monitors and glucose meters.

- **Agriculture:** M2M (Machine-to-Machine) communication has the potential to transform the agriculture industry by enabling real-time data sharing between farming equipment, sensors, and management systems. This technology can help to increase crop yields, reduce waste, and improve sustainability. One of the primary uses of M2M communication in agriculture is in precision farming. M2M communication can enable farmers to collect and analyze data from sensors and equipment, allowing them to make more informed decisions about crop management. For example, sensors can be used to monitor soil moisture levels, crop growth, and weather patterns, enabling farmers to optimize irrigation, fertilization, and other inputs to maximize crop yields. M2M communication can also be used to improve livestock management. Sensors and devices can be used to monitor the health and behavior of livestock, enabling farmers to identify potential issues early and take corrective action. For example, sensors can be used to monitor the temperature and humidity of barns and stables, alerting farmers to potential health risks before they become serious. Another area where M2M communication can be useful is in supply chain management. M2M communication can enable real-time tracking of crops, from planting to harvest to distribution. This can help to improve efficiency, reduce waste, and optimize logistics operations. M2M communication can also be used to improve sustainability in agriculture. By enabling more precise and efficient use of inputs, such as water and fertilizer, farmers can reduce waste and minimize environmental impact. Sensors and devices can also be used to monitor environmental factors, such as soil quality and biodiversity, enabling farmers to make more informed decisions about land management.

These are just a few examples of the many application areas for M2M communication. As technology continues to evolve, we will likely see even more innovative uses for M2M communication in the future.[22–25]

15.9 SERVICES FOR FUTURE M2M COMMUNICATION

As M2M (Machine-to-Machine) communication continues to evolve, there are a number of services that are likely to become increasingly important in the future:

- **Data analytics:** As the volume of data generated by M2M communication increases, there will be a growing need for tools and services that can help organizations analyze and make sense of this data. This will enable organizations to derive insights and make informed decisions based on the data generated by M2M communication.
- **Security:** As M2M communication becomes more prevalent, there will be a growing need for robust security solutions to protect data and ensure the integrity of M2M networks. This will require advanced encryption technologies, as well as other security measures to protect against cyber-attacks and other threats.
- **Edge computing:** With the increasing volume of data generated by M2M communication, there will be a growing need for edge computing technologies that can process and analyze data locally, reducing the need to transmit data back to centralized data centers. This can help to reduce latency and improve overall network performance.
- **Interoperability:** As the number of devices and systems involved in M2M communication increases, there will be a growing need for standards and protocols to ensure interoperability between different devices and systems. This will enable devices from different manufacturers and with different capabilities to communicate with each other, improving the overall effectiveness of M2M communication.
- **5G networks:** As 5G networks become more widely available, they will provide a significant boost to M2M communication by enabling faster and more reliable data transmission. This will enable a range of new M2M applications, such as autonomous vehicles, remote surgery, and industrial automation.

Inline to the services there are a number of services that will be important in the future of M2M communication, including data analytics, security, edge computing, interoperability, and 5G networks. As M2M communication continues to evolve, we are likely to see new services emerge to meet the changing needs of organizations and consumers.[24,25]

15.10 OPEN RESEARCH CHALLENGES

Despite the many advances in M2M (Machine-to-Machine) communication, there are still a number of open research challenges that need to be addressed to fully realize the potential of this technology.[25,26] Some of the key research challenges include:

- **Interoperability:** As M2M devices continue to proliferate, there is a growing need for interoperability between different devices and platforms. This requires the development of standardized protocols and interfaces that can enable devices from different manufacturers to communicate and work together seamlessly.
- **Scalability:** As the number of M2M devices and applications continues to grow, there is a need for scalable and flexible architectures that can accommodate this growth. This requires the development of new technologies and approaches that can handle the increasing volume and complexity of M2M communication.
- **Energy efficiency:** Many M2M devices are battery-powered and operate in remote or inaccessible locations. There is a need to develop energy-efficient M2M communication technologies and architectures that can enable devices to operate for extended periods without the need for frequent battery replacements or recharging.
- **Context-awareness:** M2M communication involves the exchange of data between devices and systems, and there is a need to develop technologies that can enable devices to be context-aware and adapt their behavior based on changing environmental or situational factors.
- **Real-time communication:** Many M2M applications require real-time communication between devices and systems, and there is a need to develop low-latency communication technologies and architectures that can enable real-time data exchange.

The above-discussed points are some of the key research challenges that need to be addressed to fully realize the potential of M2M communication. As researchers continue to work on these challenges, we are likely to see new and innovative M2M applications emerge that can transform industries and improve our lives.

15.11 SECURITY AND PRIVACY IMPLICATIONS

M2M (Machine-to-Machine) communication has significant security and privacy implications due to the large volume of data exchanged between devices and the potential for unauthorized access and interception. Here are some of the key security and privacy implications of M2M communication:

- **Data protection:** M2M communication involves the exchange of sensitive data, including personal and financial information. There is a need to ensure that this data is protected from unauthorized access or interception, which requires the development of advanced security and encryption technologies.
- **Authentication and access control:** M2M devices need to be authenticated and authorized before they can communicate with other devices or systems. This requires the development of authentication and access control mechanisms that can verify the identity of devices and grant access only to authorized devices.
- **Network security:** M2M communication networks need to be secure and protected from cyber-attacks, including denial of service (DoS) attacks, malware, and other forms of cyber threats. This requires the deployment of robust network security solutions, including firewalls, intrusion detection and prevention systems, and other security measures.
- **Privacy concerns:** M2M communication involves the exchange of sensitive data, including personal and location information. There is a need to ensure that this data is collected and used in a manner that respects user privacy and is compliant with data protection regulations.
- **Supply chain security:** M2M devices and systems are often manufactured by third-party vendors and suppliers. There is a need to ensure that these devices are secure and free from vulnerabilities that could be exploited by attackers. This requires the development of robust supply chain security mechanisms that can verify the integrity of devices and components.

M2M communication has significant security and privacy implications, and there is a need to develop advanced security and encryption technologies, authentication and access control mechanisms, network security solutions, and supply chain security mechanisms to ensure the safety and security of M2M networks and systems.[26–30]

15.12 CONCLUSION

Secure machine-to-machine (M2M) communication is essential for the successful implementation of 5G-enabled cities. The proliferation of connected devices and sensors in these cities creates a need for secure and reliable communication protocols to prevent cyber-attacks, data breaches, and other security threats. However, several challenges must be addressed to ensure secure M2M communication in 5G-enabled cities. These challenges include the development of robust security mechanisms and protocols that can prevent unauthorized access and protect data confidentiality, integrity, and availability. Other challenges include the need for standardized security frameworks that can be applied across different devices and systems, the management of large-scale data flows, and the integration of the existing legacy systems with new 5G-enabled technologies. Despite these challenges, secure M2M communication is crucial for the success of 5G-enabled cities. By addressing these challenges and implementing comprehensive security solutions, cities can benefit from the increased efficiency, cost savings, and improved quality of life that 5G-enabled technologies can provide.

KEYWORDS

- **M2M**
- **IoT**
- **smart cities**
- **5G**
- **privacy**
- **security**
- **integrity**

REFERENCE

1. Osseiran, A. et al. The Foundation of the Mobile and Wireless Communications System for 2020 and Beyond. *Proc. IEEE VTC—Spring Wksp.* Germany, **2013**.
2. Boccardi, F. et al. Five Disruptive Technology Directions for 5G. *IEEE Commun. Mag.* **2014,** *52* (2), 74–80.
3. Mehmood, Y. et al. Mobile M2M Communication Architectures, Upcoming Challenges, Applications, and Future Directions. *EURASIP J. Wireless Commun. Network.* **2015,** *1*.
4. ETSI, Machine-to-Machine Communications (M2M), Functional Architecture. Tech. Rep. ETSI TS 102 690 V2.1.1, Oct. **2013**.
5. Booysen, M. J.; Zeadally, S.; van Rooyen, G.-J. Survey of Media Access Control Protocols for Vehicular Ad Hoc Networks. *IET Commun.* **2011,** *5* (11), 1619–31.
6. 3GPP TSG RAN Meeting, Narrowband IoT (NB-IoT). Tech. Rep. Sept. 2015.
7. oneM2M, oneM2M Project Release 1 Specifications. http:www.onem2m.org/technical/published-documents (accessed 1 Jan. 2016).
8. T. G. I. P. P. Partnership, 5G Vision: The Next Generation of Communication Networks and Services. https://5g-ppp.eu/ (accessed 30 Dec. 2015).
9. Shariatmadari, H. et al. Machine-Type Communications: Current Status and Future Perspectives Toward 5G Systems. *IEEE Commun. Mag.* **2015,** *53* (9), 10–17.
10. Oh, C.-Y.; Hwang, D.; Lee, T.-J. Joint Access Control and Resource Allocation for Concurrent and Massive Access of M2M Devices. *IEEE Trans. Wireless Commun.* **2015,** *14* (8), 4182–4192.
11. Aijaz, A.; Aghvami, A. H. Cognitive Machine-to-Machine Communications for Internet-of-Things: A Protocol Stack Perspective. *IEEE Internet of Things J.* **2015,** *2* (2), 103–112.
12. Hasan, M.; Hossain, E.; Niyato, D. Random Access for Machine-to-Machine Communication in LTE-Advanced Networks: Issues and Approaches. *IEEE Commun. Mag.* **2013,** *51* (6), 86–93.
13. Laya, A.; Alonso, L.; Alonso-Zarate, J. Is the Random Access Channel of LTE and LTE-A Suitable for M2M Communications? A Survey of Alternatives. *IEEE Commun. Surveys Tutorials* **2014,** *16* (1), 4–16.
14. De Andrade, T. P.; Astudillo, C. A.; da Fonseca, N. L. Allocation of Control Resources for Machine-to-Machine and Human to-Human Communications Over LTE/LTE—A Networks. *IEEE Internet of Things J.* **2016,** *3* (3), 366–77.
15. Priyanka, K.; Mallavaram, G.; Raj, A.; Pradhan, D. Cognitiveness of 5G Technology Toward Sustainable Development of Smart Cities. In: *Decision Support Systems for Smart City Applications*, **2022**; pp 189–203.
16. Pradhan, D.; Ray, S.; Dash, A. A Critical Review on Sustainable Development of Green Smart Cities (GSCs) for Urbanization. *J. Altern. Renew. Energy Sources* **2022,** *8* (3), 21–29. https://doi.org/10.46610/joares.2022.v08i03.003
17. Pradhan, D.; Dash, A.; Tun, H. M.; Wah, N. K. S.; Oo, T. A Sustainable Key Enabler for mm-Wave Beamforming in 5G Environment. *J. VLSI Design Signal Process.* **2022,** *8* (3), 10–17. https://doi.org/10.46610/jovdsp.2022.v08i03.002
18. Pradhan, D.; Tun, H. M. Security Challenges: M2M Communication in IoT. *J. Electr. Eng. Autom.* **2022,** *4* (3), 187–199.

19. Pradhan, D.; Tun, H. M.; Dash, A. K. IoT: Security & Challenges of 5G Network in Smart Cities. *Asian J. Convergence Technol. (AJCT)* **2022,** *8* (2), 45–50.
20. Pradhan, D.; Dash, A.; Tun, H. M.; Wah, N. K. S.; Oo, T. Improvement of Capacity and QoE: Distributed Massive MIMO (DM-MIMO) Technology-5G. *J. Netw. Secur. Comput. Netw.* **2022,** *8* (3), 9–17. https://doi.org/10.46610/jonscn.2022.v08i03.002
21. Alshenaifi, I. M.; Qazi, E. U. H.; Almorjan, A. IoT Forensics: Machine to Machine Embedded with SIM Card. In: *ITNG 2023 20th International Conference on Information Technology-New Generations*; Springer International Publishing: Cham, **2023**; pp 133–142.
22. Noman, S. A.; Noman, H. A.; Al-Maatouk, Q.; Atkison, T. Internet of Things Communication, Networking, and Security: A Survey. *Int. J. Comput. Digit. Syst.* **2023.**
23. Ihita, G. V.; Acharya, V. K.; Kanigolla, L.; Chaudhari, S.; Monteil, T. Security for oneM2M-Based Smart City Network: An OM2M Implementation. In: *2023 15th International Conference on COMmunication Systems & NETworkS (COMSNETS)*; IEEE, **2023**; pp 808–813.
24. Noor, A.; Ratul, M. S.; Ahmed, A. I.; Hassain, H.; Ahmed, A. An IoT Based Smart Grid: Peer-to-Peer Energy Trading for Electric Vehicles Using M2M Communication Technology. In: *2023 3rd International Conference on Robotics, Electrical and Signal Processing Techniques (ICREST)*; IEEE, **2023**; pp 289–293.
25. Singh, H. W.; Devaki, K.; Fernandes, J. V.; Pradhan, D. PoD Vs SNR Estimation: C-MIMO Radar Using STC and STAP Algorithm. In: *2022 IEEE Region 10 Symposium (TENSYMP)*; IEEE, 2022; pp. 1–6.
26. Pradhan, D.; Sahu, P. K.; Tun, H. M. A Study of Localization in 5G Green Network (5G-GN) for Futuristic Cellular Communication. In: *Proceedings of the 3rd International Conference on Communication, Devices and Computing: ICCDC 2021*; Springer: Singapore, 2022; pp 453–465.
27. Pradhan, D.; Priyanka, K. C. RF-Energy Harvesting (RF-EH) for Sustainable Ultra Dense Green Network (SUDGN) in 5G Green Communication. *Saudi J. Eng. Technol.* **2020,** *5* (6), 258–264.
28. Bhat, K. U.; Kumar, N.; Koul, N.; Verma, C.; Enescu, F. M.; Raboaca, M. S. Intelligent Communication for Internet of Things (IoRT). In: *Proceedings of International Conference on Recent Innovations in Computing: ICRIC* **2022,** *Volume 2*; Springer Nature: Singapore, 2023; pp 313–328.
29. Yuan, F.; Zhang, Y.; Zhang, J. IoT Technology for Intelligent Management of Energy, Equipment and Security in Smart House. *Int. J. Adv. Comput. Sci. App.* **2023,** *14* (1).
30. Rao, P. M.; Deebak, B. D. A Comprehensive Survey on Authentication and Secure Key Management in Internet of Things: Challenges, Countermeasures, and Future Directions. *Ad Hoc Netw.* **2023,** 103159.

Index

H

I

K

M